TORONTO THE WILD

TORONTO THE WILD

FIELD NOTES OF AN URBAN NATURALIST

WAYNE GRADY

MACFARLANE WALTER & ROSS

TORONTO

Macfarlane Walter & Ross
37A Hazelton Avenue
Toronto Canada M5R 2E3

Canadian Cataloguing in Publication Data

Grady, Wayne
Toronto the wild

Includes bibliographical references and index.
ISBN 0-921912-90-0

1. Urban ecology (Biology) – Ontario – Toronto.
2. Urban fauna – Ontario – Toronto.
3. Urban flora – Ontario – Toronto.   I. Title.

QH106.2.05G73 1995   574.5'268'09713541
C95-932171-3

The publisher gratefully acknowledges the support of the
Canada Council and the Ontario Arts Council

Printed and bound in Canada

For Merilyn, with whom these walks
and so much else were made

CONTENTS

ACKNOWLEDGEMENTS

These essays were written after long talks with many experts, without whom I wouldn't sound nearly as authoritative as I have tried to be. Those who were especially generous with their time and knowledge include Rick Rosatte and Dennis Voigt at the Ontario Ministry of Natural Resources; Bob Johnson and Diane Devison at the Metro Toronto Zoo; Tim Myles, director of Urban Entomology at the University of Toronto; A.E.R. Downe, who spoke to me at Queen's University in Kingston; and Hans Blokpoel of the Canadian Wildlife Service, Ontario Division. Sandra Zelmer, at the Toronto Animal Control Office, was also very helpful. Wendy Hunter and Paloma Plant, at the Toronto Humane Society's Wildlife Department, and Nathalie Karvonen, at the Toronto Wildlife Centre, gave me many hours of their frenzied time. Fred Bodsworth, David Macfarlane, Graeme Gibson, Brian Fawcett and M.T. Kelly shared their experience and their enthusiasms with me; Geri Clever went them one better and gave me tea and pound cake.

I also want to thank Ed Thomas and Wendy Thomas (no relation) for their invaluable help in the research and editing of these essays.

"The Mammal Behind the Mask" and "The Post-Colonial Termite" first appeared, in somewhat different forms, in Toronto Life magazine.

The lines from James Reaney's poem A Suit of Nettles, which appear on page 206, are reproduced with the permission of Porcépic Books, an imprint of Beach Holme Publishers

Some financial assistance was provided by the Ontario Arts Council, for which I am very grateful.

INTRODUCTION

"It is usually supposed," the celebrated British naturalist Richard Jefferies wrote in *Nature Near London*, published in 1893, "to be necessary to go far into the country to find wild birds and animals in sufficient numbers to be pleasantly studied." That supposition has become even more usual in the intervening decades. From the point of view of nature lovers, cities are dead piles, blights on the natural landscape. "The word 'nature'," writes Charles Simic, an American novelist, "has always had for me, a city boy, unpleasant didactic connotations. It was the place our mothers and our schools took us occasionally so we could get some fresh air." Even city-dwellers who actually *like* nature are conditioned to look for it outside the city: at the cottage, in the park, at summer camp, up north. Somewhere else. "For those living in cities," writes Moira Farr in a recent issue of *Brick* magazine, "'nature' tends to be a separate thing, 'out there.' The Great Outdoors. The Unspoiled Wilderness. The place you go when you need to 'get away from it all.'" This "nature" is always referred to in inverted commas, as when talking about something ersatz — like calling an orange-coloured, processed, milk by-product "cheese." Not real nature, but "nature" as an earthly paradise, a pastorale, a green thought in a green shade, far removed from the urban sprawl. Nature as idea.

There is also Northrop Frye's idea of nature. Frye con-
ceived of nature as the opposite of civilization, as the barbar-
ian waiting at the gates, a threat to physical comfort and
social stability. "Anything that isn't middle class," he once told
me in his book-lined office at the University of Toronto, "is
still in the trees." Frye imagined European settlers sailing into
the mouth of the Gulf of St. Lawrence "like Jonah being swallowed by the whale" and built this conception of nature as devourer into a theory of Canadian literary consciousness. Fear
of nature, he said, has caused Canadians to develop "a garrison
mentality": we build walls around our cities to keep nature out
(or around our zoos, to keep nature in). And we venture beyond those walls only in large groups, armed to the teeth,
peering nervously over our shoulders. Nature as nightmare.

I'm afraid I don't share either view. Frye's idea of nature has
some weight as long as he confines himself to writers of fiction and poetry, but Canada's explorers and nature writers did
not regard the wilderness as an enemy, but rather approached
it and traversed it with an almost religious sense of wonder.
Susanna Moodie might have portrayed "the bush" as a rough
and threatening place, but her sister, Catherine Parr Traill,
was an accomplished and respected naturalist, whose *Studies of
Plant Life in Canada* (1885) was highly regarded by professional
botanists the world over. And although, like Charles Simic, I
am a city boy, I don't recall my parents or my teachers taking
me out of the city to get fresh air. I was born in Windsor,
Ontario, and there are family stories about swimming trips to
Point Pelee, and visiting Jack Miner, the famous bird conser-
vationist, and of me standing beside Miner patting a Canada
goose, but I don't remember any of that. What I remember is
playing in an open field across the street from our house on
Factoria Avenue, whole summers spent lying in tall grass
beside a stream that ran through the field, watching
grasshoppers and crickets, tadpoles and garter snakes. I blew
the heads off dandelions, checked to see if my mother liked

butter by holding a buttercup under her chin (she always did), and punched air-holes in the lids of jam jars so I could collect the chrysalids of monarch butterflies and watch them hatch. I do not say that I was a budding field naturalist – I was no more curious or knowledgeable about nature than any other child of seven or eight – but it is true that when my family moved out of Windsor into a farming community near Barrie, Ontario, when I was nine, the transition from city to country was not, for me, as traumatic as it might have been had I not had the field.

Canadians are not afraid of nature, and cities are not garrisons; or at least if they are, they are highly ineffectual ones, for there is as much of nature in the city as there is out of it. So much, in fact, that it becomes difficult to think of the city as being somehow separate from nature. "The city," writes physicist James Trefil, "can be thought of as a natural system. . . . At the most obvious level, a city is an ecosystem, much as a salt marsh or a forest is. A city operates in pretty much the same way as any other ecosystem, with its own peculiar collection of flora and fauna." Cities are, after all, built directly on top of wildernesses, and they don't always displace the plants and animals that live there. A city is not an artificial construct, superimposed on a natural landscape: it is part of the landscape it inhabits. "It seems to me," continues Trefil, "that we should begin our discussion of cities by recognizing that they aren't unnatural, any more than beaver dams or anthills are unnatural. Beavers, ants and human beings are all products of evolution, part of the web of life that exists on our planet."

Toronto has been built in a particularly opportune spot, from a naturalist's point of view. It is often said that something like half of Ontario's prime farmland can be seen from the top of the CN Tower, and prime farmland – whether farmed or not – supports an enormous range of plants, shrubs, trees, and the myriad birds and animals that go with them. Toronto also happens to be located at precisely the spot where two major bird-migration flyways intersect, which means that in the spring and fall the city is visited by or home to something like 304 species of birds. The city itself occupies the fertile flood-plains of three significant rivers – the Don, the Humber, and the Rouge – and ten creeks, each of which has carved deep valleys into the sandy loam that has accumulated between the Niagara Escarpment and the shore of Lake Ontario. In 1906, James H. Fleming, one of Toronto's best known ornithologists, described the city from the point of view of a naturalist:

> The city for a greater part of its width is protected from the lake by a sandbar and island, once continuous. The sandbar runs west from near the eastern city limits for nearly three miles till it is divided by the Eastern Channel, and sending a spur north encloses what is known as Ashbridge's Bay. This is really a marshy lagoon of considerable size, and though filled in, in places, still affords food and shelter for many species of birds. Into this bay originally drained some eleven creeks, and at its western end the River Don, which now is confined to an artificial channel and flows into Toronto Bay somewhat further north than where the original outlet of Ashbridge's Bay was.

Although Ashbridge's Bay is now greatly diminished by the diversion of the mouth of the Don River – Toronto's restaurants now have to get their ducks from farther afield – that part of Toronto Harbour still affords food and shelter for many birds and other forms of wildlife. It is now the site of Leslie Street Spit, a human construct composed entirely of rubble from Toronto's wrecking frenzy of the 1970s, when most of the downtown core's venerable buildings were torn down to make room for such architectural innovations as the Eaton Centre and the Royal Trust Tower. But the rubble has

been our ornithological gain, for the Spit, which curves out
five kilometres into the lake following almost exactly the con-
tours of the sandbar that formerly contained Ashbridge's Bay,
is the site of what is probably the largest ring-billed gull
colony in the world. Altogether, 46 species of wildfowl winter on the Spit, and more than 258 species of birds have been
recorded on it, including Caspian terns, black-crowned night
herons, blue-winged teals, Hudsonian godwits, American
kestrels, northern harriers, falcons and 34 different species of
plovers. Nor are the many attractions of the Spit – open to
the public only on weekends, when it is better known to naturalists and in-line skaters as Tommy Thompson Park – confined to birds. Tommy Thompson was Toronto's parks
commissioner in the 1960s and 70s. (It was Thompson who
coined the phrase that is now the parks commission's motto:
"Please walk on the grass.") A survey conducted in the 1980s
found that the Spit had also been colonized by no fewer than
283 species of wild plants, from white sweet clover (*Melilotus
alba*) to the rare sand spurrey (*Spergularia marina*), found
nowhere else in Ontario except on the shores of James Bay.
There is a colony of entirely black garter snakes, very rare,
somewhere on the Spit, though I have yet to locate it. And
during my frequent excursions onto the peninsula in all seasons, I have come across the tracks and sign of numerous
mammals, including weasels, muskrats, grey squirrels, snow-
shoe hares, and even coyotes.

> Toronto had originally many small ravines [Fleming con-
> tinues], through which flowed the streams that emptied
> into the water front. Most of these ravines are now filled
> in; in the northeast part of the city, in what is known as
> Rosedale, ravines of considerable depth exist and cross
> the back of the city to the valley of the Don; to the west of
> the city the ravines are not so numerous, though there are
> several between the western city limits and the Humber.

This river and the Don run for some distance through flats between high banks.

Many more of Toronto's ravines were filled in after the ravages of Hurricane Hazel, which swept through the city in 1954, tearing up trees and causing great erosive water damage. When the storm abated, storm drains were installed and buried in the valleys in anticipation of fresh disasters, which never came. But there are enough watercourses left to provide Toronto with some vestiges of its original wildness, especially when combined with the city's many parks, zoos, public gardens, cemeteries, and private grounds, some landscaped, some not. Forested ravines still radiate from the Don and Rouge rivers. There are kilometres of trails and footpaths throughout the city, and literally dozens of specially demarcated nature areas, from Highland Creek Park in the east to Sawmill Creek in Mississauga, each with its own particular flora and fauna and purpose for being preserved. That view from the top of the CN Tower is of an almost unbroken expanse of green foliage, so that it is difficult to discern where is forest, and where is not.

A walk along the Humber River, from Bloor south to Lake Ontario, can be an excursion into a Toronto not much different from the days when the Humber was called St. John's Creek. In 1793, Mrs. John Graves Simcoe travelled to the creek from her residence in York:

> Wed. 4th Sept.: I rode to St. John's Creek. There is a ridge of land extending near a mile beyond St. John's House, 300 feet high and not more than three feet wide; the bank towards the river is of smooth turf. There is a great deal of hemlock spruce on the river; the banks are dry and very pleasant. I gathered a beautiful large species of Polygaia, which is a genus of annual and perennial herbs and shrubs of the order of Polygalacae. I found a green caterpillar,

with tufts like fir [sic] on its back. I accidentally touched
my face with them, and it felt as if stung by a nettle, and
the sensation continued painful for some time.

The Polygalaceae is the milkwort family (*gala* is the Greek
word for milk), and Mrs. Simcoe was probably describing the
Seneca snakeroot (*Polygala senega*), a tallish stalk with white or
pinkish flowers that thrives in sterile soil (William Scott, in
"The Seed Plants of Ontario" – his chapter in *The Natural History of the Toronto Region* published in 1913 – describes the soil
of Toronto as all clay and sand in the valleys, and between
them, "cold bogs and swamps"). Calling milkwort a "herb" is
unusual: Seneca snakeroot was used by the Senecas as a cure
for snake bite, and Mrs. Simcoe frequently refers to the abundance of massasauga rattlers (which are in fact not found anywhere near Toronto). But the plant was also discovered in
1736, by the Virginian doctor John Tennent, to be useful in
the treatment of pleurisy and pneumonia: was Mrs. Simcoe
perhaps suffering from a summer cold?

The day in June when we walked down the Humber, Bloor
Street above the valley was as hot and dusty as only full sun on
concrete can be, yet in the tangled Carolinian underbrush the
sunlight was stippled and the air cool and hushed. The ground
was white with cottonwood cotton. We saw no milkworts, but
there was plenty of dame's rocket (*Hesperis matronalis*), a phlox-
like garden escapee that would have surprised and delighted
Mrs. Simcoe. We also saw cow parsnips (*Heracleum maximum*),
their huge leaves patterned by leaf miners; dainty Canada
anemones (*Anemone canadensis*); daisy fleabanes (*Erigoreron an-
nuus*); and purple-flowering raspberries (*Rubus odoratus*), with
their pink, rose-like flowers, maple-shaped leaves, and tart
berries. In the damp shade of the cottonwoods, under twisted
arbours of wild grape and Virginia creeper, the river was quiet
and sun-dappled. Canada geese and mallard ducks paddled in
it, and white-throated sparrows held the higher branches of

the cottonwoods. We didn't see Mrs. Simcoe's stinging caterpillars, but their role had been adequately taken up by vast
stretches of stinging nettles (*Urtica dioica*), another introduced
plant, which made our bare legs continue painful for quite
some time indeed.

As mention of those introduced species suggests, there is
actually more nature in Toronto now than there used to be. A
city attracts wildlife to it that otherwise wouldn't be found
within hundreds of kilometres of the place. Of the 283
species of vascular plants on Leslie Street Spit, only 115 are
native to the Toronto area. Oddly enough, that percentage
decreases downtown: one vacant lot examined in the Annex,
between Spadina and Bathurst north of Bloor, contained
thirty-two species of wildflowers, and only seven were native
– goldenrod, ragweed, wood sorrel, evening primrose, common cinquefoil, wild oats, and milkweed. When L. T. Owens
presented his paper, "Trees of the Toronto Region," to the
405th meeting of the Brodie Club, held at the Royal Ontario
Museum in 1944, he listed 215 species, "fifty-one of them
indigenous." A great many of the plants that we now consider
to be synonymous with Toronto summers are in fact invaders
(or invitees) from another world – the Old World – part of
what agricultural historians call the Columbian Exchange,
since it began with Christopher Columbus bringing maize
and sweet potatoes back to Europe and introducing cucum-
bers, radishes, and parsley to the West Indies. He truly believed he had found the Garden of Eden in the New World,
and he thought it needed cultivating. We have since continued the cultural exchange, sometimes deliberately, often
inadvertently, and the result has been a net increase in the
number of species that can be found by even a casual observer
of Toronto's wildlife.

    The dozens of animals, birds, plants, and insects discussed
in these essays are a case in point. Some of them were here

before there was a place called Toronto (which was, by the way, originally the Mississaugas' name for Lake Simcoe): Europeans arrived to find mosquitoes, raccoons, skunks, and snakes already in residence and reluctant to move out, and we have since established an uneasy rapport with most of them. Other species moved in because the houses and landscapes we created for ourselves also looked pretty good to them. Big brown bats wouldn't winter in the city if we hadn't built attics specifically, so it must seem to them, for their use. Coyotes were not found in Ontario at all until the 1940s, when they migrated east to exploit the large tracts of rodent-rich grasslands that we call suburban parkland, but that so resemble the coyote's native Great Plains. There probably wouldn't be cockroaches in the New World if we hadn't brought them from Africa on slave ships trading in the West Indies, and even then they probably wouldn't have spread this far north if we hadn't invented central heating. The same goes for termites: we may be witnessing the evolution of a whole new species of subterranean termites in Toronto, a direct result of the old species' ability to over-winter in the cold, Canadian ground now that we have warmed it up with heated sewer and subway lines and underground power cables. Crows and some gulls now over-winter in Toronto because our effluents never freeze. We have altered our environment to suit ourselves, and we have found to our amazement and sometimes consternation that this new world also suits a few uninvited species.

There have, of course, been some deliberate introductions, although not as many as the accidental kind. Of the sixty-five alien plant species on Leslie Street Spit, only six are garden escapees. The starling, the house sparrow, the pigeon, and the house finch have all been brought to North America for various and often spurious reasons. Not all introductions have been disastrous: ox-eye daisies (*Chrysanthemum leucanthemum*), for example, a European flower that, though listed in the 1911 Ontario Department of Agriculture's *Weeds of Ontario* as a noxious weed ("most troublesome in pastures, and can be got rid of only by cultivation"), has earned a permanent place in our urban cosmology. As have lilacs. Native to Eurasia, lilacs (*Syringa vulgaris*) must now be the most familiar free flowers on Earth. Members of the family Oleaceae, a universal family that includes forsythia (*Forsythia suspensa*), jasmine (*Jasminum officinale*), and white, red and black ash (*Fraxinus americana, pennsylvanica* and *nigra*), lilacs are distantly related to the olive – *Olea europaea.* The lilac was brought to Europe in 1562 by Ogier Ghiselin de Busbecq, the German ambassador to Constantinople, who thought he would brighten up the bleak German landscape with a whiff of the exotic Persia with which he had fallen in love. The word "lilac" is our version of the Persian *nilak,* which means "dark bluish." Lilacs came to Canada with the colonists – those who contend that Ontario's earliest settlers were too busy scraping a living out of the soil to bother with aesthetics have trouble explaining lilacs, which for one week in May are beautiful and aromatic and entirely useless in any way related to the Protestant work ethic, and yet were found in every country garden. Abandoned homestead sites in rural Ontario can still be found simply by looking for lilacs – find a clump of lilac bushes, and there will be an old stone cellar somewhere in the vicinity, full of old ketchup bottles. And if you want to know where the outhouse was, look for the rhubarb. I have always liked the association of rhubarb with outhouses, because it reminds me that rhubarb was first imported into England (from China) by the East India Company in the 1740s, not for its stalk but for its root, which, dried, powdered, and made into pills, was a natural laxative.

A word here on the ordering of the essays in this book. It is vaguely seasonal. Dividing a year into four distinct seasons is natural enough at these latitudes, but placing those divisions precisely on specific days seems to me to be obsessively tidy-minded. Summer, for instance, does not suddenly happen on

June 21, solstice or no solstice. In the city, the arrival of summer is more a matter of psychology than astronomy. We know it's here when we don't automatically reach for our jacket when we go outdoors. When we think of restaurants we think of sidewalk cafés, and not just the ones on the sunny side of the street. Window screens enter our consciousness (are they in the basement or under the spare bed?). Baseball begins to make more sense than hockey. Bicycles actually get the oil they've been squeaking for for weeks. The empty propane tank in the garden shed or out on the balcony may be hauled out, wiped clean of cobwebs and dust, and left in a conspicuous place as a kind of mnemonic device or talisman. But *when* these changes take place varies from person to person and from year to year.

The essays in this book are arranged in a way that reflects these changes. By beginning with the first essay and progressing through to the last, the reader will have traversed a year of the bird, plant, and animal lives that share our urban habitat, in more or less the correct order but with a few overlaps and surprises. I think of starlings as winter birds, but of course they are here all year. Snakes come out of hibernation in the spring, mosquitoes make themselves most noticeable in the summer, and bats come home to roost in the fall. But do not look for sharp demarcations between these natural events, because there aren't any, neither here nor in nature.

> Returning again to the city, [continued J.H. Fleming in 1906], the land rises gradually from the water front for some two and a half miles, and at North Toronto is 160 feet above the lake. From here an ancient lake margin rises abruptly some 70 feet to a plateau which sweeps across the back of the city and is broken only by the valley of the Don on the east, and the Humber on the west, and a few small ravines; a good deal of wood remains along this rise. This ancient water margin is one of a number (said to be

thirteen) that exist between here and Lake Simcoe, some
60 miles further north; the highest point, 26 miles north
of the city, near King, is 780 feet above Lake Ontario. . . .

The geology of Toronto hasn't changed perceptibly in at least
10,000 years. Once the weight of the glaciers receded, and
Iroquois Lake drained off into the Gulf of St. Lawrence,
southwestern Ontario began rebounding at the rate of a few
millimetres per century. This can be startling to some. At the
May 1900 meeting of the Hamilton Naturalist's Association,
Colonel C.C. Grant alarmed the audience during his opening
address by announcing that "recent observations prove that a
large part of the region around the Great Lakes of North
America is being raised or lowered in consequence of the
Earth's internal forces." And it is still fascinating to think that,
not so long ago, Point Pelee was the western end of Lake Erie.
I find such gradual change comforting. I rather like the idea
that if I stand in one place long enough, I will eventually be in
a different place. Nature loves change, which is why it abhors
a vacuum. In a way, that is how I meant these essays to work; I
would like reading them to be like following a meandering
stream as it gurgles pleasantly through woodland and pastures, sidles past an old mill, ripples under a modern highway
overpass, and drains into a reservoir. In contemplating a lilac
bush or a cockroach, we have travelled from Constantinople
to sixteenth-century Germany, from pre-contact Africa to the
West Indies, and back, always back. To watch a dandelion
head open and turn to the sun, or a pigeon pecking at grass
seeds in the park, is to experience in one minute the history of
life on this planet. And perhaps by realizing that neither the
dandelion nor the pigeon nor the grass would be there were it
not for us, we become aware of our own place in the great
web of life.

Spring

I, the invincible,
    March, the earth-shaker;
March, the sea-lifter;
    March, the sky-render.

— *Isabella Valency Crawford*

Make me over, Mother April,
When the sap begins to stir!
Make me man or make me woman,
Make me oaf or ape of human,
Cup of flower or cone of fir;
Make me anything but neuter
When the sap begins to stir!

— *Bliss Carman*

THE MAMMAL
BEHIND THE MASK

Here it is the middle of March, and the three raccoons that spent most of last summer in the maple tree at the end of our yard haven't come back. I used to go out in the evenings to watch them work their way down the trunk, a large female and two wide-eyed adolescents, she first, tentative, testing the ground, then the two insouciant kids. They'd join up at the base of the tree and file off along the alley, checking out the garbage pails and compost heaps as they went, until they disappeared into the twilight. One night we had a corn roast in our back yard and they didn't come down, but stayed in the tree looking down at us, maybe smelling the corn, their three intelligent faces watching the party like neighbours waiting for an invitation.

In the mornings they'd return, usually from a different direction, the female waddling along in her odd, hump-backed fashion, as though she had been designed to walk upright but had been forced down onto all fours and was not liking it. Then the young ones, tired but still playful, still looking around. Once in a while they would surprise the neighbour's cat, who would be sitting in our garden eyeing the nest that a pair of house finches had made in our ivy; she'd freeze as the raccoons paraded past her, then scamper off

home to her Puss'n'Boots. The coons would climb back up the
tree, take their positions in three forks that held them like
cupped hands, and sleep through the day. I miss them.

I suppose I did regard them as neighbours. They certainly
weren't pets. Raccoons do, in fact, take very good care of
themselves. There are an estimated 40,000 of them living in
Toronto. In some places – along the Don Valley, up the
ravines, anywhere where there is good tree cover – they
achieve a population density of up to 100 per square kilometre, compared to the rural average in Ontario of about four.
Toronto has been called the Raccoon Capital of North America. In Scarborough, where the Ministry of Natural Resources
(MNR) has been vaccinating raccoons and skunks since 1987
in an effort to curb rabies, the density is about sixteen per
square kilometre. Any way you look at it, raccoons have been
very successful city residents.

"Here you go, guys. Come and get it."

Geri Clever (pronounced Cleaver) and I think alike when
it comes to raccoons. At the moment, she's standing at her
back door and shaking a plastic scoopful of No Name Dog
Kibble out onto the driveway. Milling around at her feet
are half a dozen black-masked banditos, looking up at her
while they feel around on the asphalt with their uncannily
humanoid hands for the kibble. Raccoons do not pick up
their food with their teeth, as cats and dogs do, but take it up
delicately with their fingers and roll it between their black
palms to soften it before popping it into their mouths and
chewing it in studied concentration, like connoisseurs trying
to identify a subtle aftertaste. They do not, however, take
their eyes off Mrs. Clever. Once in a while, hands from two
different raccoons fall on the same piece of kibble, and a brief
snarling match ensues, which reminds us that these are still
wild animals we are feeding, but by and large it is a peaceful
gathering.

"There are usually four or five more," says Mrs. Clever, peering off into the darkening ravine behind her house. It's 7:30 p.m., and the still leafless trees below the lip of the ravine are quickly losing definition. Mounted on the outside door-post is a small brass bell, like a miniature ship's bell, with a cord dangling from the clapper. When she yanks the cord, the sound of the bell fills the night, and within minutes three more pairs of eyes peer over the lip of the ravine. "Here they are," she says cheerily. She shakes out another scoopful of kibble from a red plastic pail beside the sofa and then slides the door closed. "It used to be an ordinary door with a handle," she says, crossing the room, "but they soon figured out how to work it, and they'd just saunter in at all hours of the night to raid the cat's dish in the kitchen. They even knew which closet we kept the kibble in. So five years ago we had it changed to a slider."

She sits down at the glass-topped table in the sunroom, where we're drinking coffee and eating pound cake, and resumes flipping through her file of photocopied newspaper clippings and magazine articles about raccoons and rabies. There are also stacks and stacks of photographs, all of raccoons – fat raccoons, ragged-looking raccoons, raccoons with no tails, raccoons in trees. It's a thick file, thicker than mine. But then she's been at it longer. She is Toronto's most enthusiastic raccoon supporter. She's been feeding raccoons from her back door every night for the past twenty-six years. "We started out just giving them bread," she says. "We'd go to a bakery every day and buy bags of day-old white bread. But that got too expensive – you know, back then, we were feeding up to thirty-two raccoons a night, and they'd eat us out of house and home if we let them." Back then was 1967, when she and her husband first bought the house on Rosedale's Beaumont Road. It's nicely situated on a long, narrow ridge poking into the Don Valley: looking out through the walls of her sunroom, I can see nothing but a dark and quiet wilderness. "So we switched to dog kibble. Even that was expensive, until No Name came along. They'd go through a fifty-pound bag in a couple of weeks if we let them. Thank God for Dave Nichol, I say," and she laughs.

Outside the sliding glass door, three raccoons raise themselves up on their haunches and peer in at us, shielding their eyes with their hands – a male and two females. "The only time I can tell which is which," she says, "is when they're standing up like that."

Like all city-dwellers, including humans, raccoons are omnivorous, adaptable, and smart. In the wild, they eat anything from crayfish, earthworms, and frogs to acorns, berries, and fruit. They catch fish and do not scorn small mammals. They will get into a henhouse and kill chickens, and they have been known to bring down sheep. In the city, where the food supply is virtually unlimited, the word omnivore takes on a whole new meaning. The list of food items found in the stomachs of urban raccoons reads like a pig-out at an A&P after a weekend at the CNE: bread, candy, bird seed, eggs, hot-dogs. "They'll eat anything but onions," says Mrs. Clever. "And maybe celery. They don't seem to like greens very much. They love spaghetti. And corn, of course – but I have to cook it for them, they won't touch raw corn."

Mrs. Clever puts a dish of water out with the kibble, but the raccoons hardly touch it. The idea that they have to wash their food in water before eating it is a myth. A very powerful myth, to be sure, and one that prompted Carl Linnaeus to give them the scientific name *Ursus lotor*, which means "washing bear." They sometimes swish food around in water, if there is some about, but they don't have to. Their favourite foods are aquatic – crayfish, snails, and such – and they spend a lot of time at night feeling around on the bottoms of streams and ponds for these items, and the myth may have grown out of that. But I can't imagine a wild raccoon dragging a sheep to a nearby stream in order to wash it. Captive raccoons seem to wash food more often than wild raccoons, and some biologists theorize that they do this to give themselves the illusion that they have caught the food item themselves; if that is so, then my estimation of their intelligence goes way up. Other researchers claim that wetting increases the palms' sensitivity, making it easier for raccoons to distinguish food items from lumps of mud. Both theories are improvements over the old idea that raccoons had to wet their food because they didn't have any salivary glands. If they didn't produce saliva they wouldn't be able to transmit rabies, and we know they can transmit rabies.

Sometime in the mid-1970s, raccoon hunters in West Virginia ran out of raccoons on which to train their coon hounds. So, no problem, they imported a bunch of new ones from Florida. They brought in several lots, totalling about 3,500 raccoons, and turned them loose in areas managed by private hunt clubs. As anyone who has been to Florida can attest, its peninsular shape makes it different from the rest of the continent. Separated from the mainland by a narrow swath of dry savannah, its swampy tip harbours wildlife not found elsewhere in North America (birds, for example, like the smooth-billed ani and the crested caracara), and some of the fauna that *is* found elsewhere has developed a few peculiarities unknown north of the Sunshine State. The raccoons imported by the West Virginia hunters looked like ordinary raccoons, but in one significant respect they differed from raccoons found anywhere else on the continent: many of them had raccoon rabies.

When a rabid raccoon was taken near the border between Virginia and West Virginia in August of 1977, alarm bells didn't immediately ring, even when it was revealed that the critter had raccoon rabies. Rabies in wild animals was not a new phenomenon, after all, and though it is known to be a horrific disease when contracted by humans, modern medical science has pretty much kept it under control in North America. Worldwide, about 25,000 people a year die from rabies, but most of them are in Third World countries, and they get it from mad dogs: in the United States and Canada, where vaccinating pets against rabies is common, the death figure is about one per year. For the twenty years before 1977, the closest most of us had come to any kind of rabies was watching reruns of *Old Yeller* on Sunday-afternoon TV. All that was about to change, but no one realized it at the time.

Rabies is a virus that attacks the central nervous system of vertebrates, particularly canines. In fact, it was once thought to be exclusively transmitted to other mammals by mad dogs. The Roman encyclopedist Aulus Cornelius Celsus – from whose first-century treatise on medicine we have adopted such Latin anatomical terms as abdomen, tonsil, vertebra, anus, and uterus – prescribed treating *rabere* ("to rage") by ducking the victim in cold water, excising the wound with a scalpel, and cauterizing the tissue with a red-hot iron, a treatment that was followed with indifferent success for the next seventeen centuries, until Louis Pasteur developed a vaccine for the disease in 1885.

It wasn't until twenty years ago that medical scientists, with the help of the electron microscope, were able to isolate and characterize the rabies virus itself. Until then, it was thought that all rabies virus strains were identical. In other words, that rabies in all animals susceptible to the disease – dogs, cats, foxes, jackals, wolves, raccoons, yellow mongooses, skunks, bats, and cattle – was one kind of rabies. We now know that each species of animal carries its own strain. There is canine rabies, sylvine rabies, bovine rabies, equine rabies, skunk rabies, and so on. Any mammal can contract any strain – a horse can get fox rabies, and skunk rabies, transmitted through pets, accounts for more than half of the 30,000 humans treated for rabies in North America annually. The rabid raccoon examined in West Virginia in 1977 had raccoon rabies. Usually, rabies is transmitted in the saliva, through bites, but in some cases the virus can be transmitted through air. Coyotes placed in bat-proof cages in a cave in the southern United States all contracted bat rabies and died without having touched an actual bat. There is a lot about rabies we don't know.

What we do know isn't pleasant. Once the virus enters the victim's bloodstream, it penetrates the nerve endings and migrates to the spinal column at the rate of about three millimetres per hour. When it hits the spinal cord, it travels slowly upward to the brain, where it rapidly colonizes and multiplies, shooting back along nerve trunks to other parts of the body, becoming especially virulent in the salivary glands. In the victim, the first signs of infection are an alert, slightly troubled demeanour, as though the intelligence has become aware that something is wrong but hasn't quite figured out what it is. This stage is followed by a ceaseless, unsatisfying restlessness, with the victim snapping at anything that moves. "The patient can neither stand nor lie down," wrote the Italian physician Girolamo Fracastoro in 1546. "Like a madman, he flings himself hither and thither, tears his flesh with his hands, and feels intolerable thirst." It is this thirst that has given human rabies the alternative name "hydrophobia," for the victim becomes dehydrated and at the same time is terrified of water, especially moving water. Constriction in the throat prevents swallowing; the lower jaw hangs open, saliva drools down the chin. "It is then," concludes Fracastoro, "that they bite other persons, foam at the mouth, their eyes look twisted, and finally they are exhausted and painfully breathe their last."

Until 1977, raccoon rabies had never been found outside Florida. Which explains why, until it began showing up in West Virginia, no one realized that the alarming thing about raccoon rabies was that it didn't respond to any of the treatments that were used to deal with other kinds of rabies. It was a unique strain. And when it began spreading north, rapidly, health officials found themselves unable to stop it.

When Peter Kalm, one of Linnaeus's students, travelled to North America in 1749, he observed in the wild much, including raccoons, that his master had seen only in captivity or had read about in reports. The raccoon, Kalm wrote in 1770, "is found very frequently and destroys many chickens. It is hunted by dogs, and when it runs up a tree to save itself a man climbs up after it and shakes it down to the ground, where the dogs kill it. . . . The bone of its male parts is used for a pipe cleaner. The hatters purchase their skins and make hats of them, which are next in quality to those of beavers. The tail is worn round the neck in winter and therefore is likewise valuable. The flesh," he noted, "is eaten and reputed to taste well."

Mrs. John Graves Simcoe, wife of Upper Canada's first governor-general (and the founder of York), noted in her diary on November 20, 1793, that after spending the afternoon in the woods on what is now Toronto Island, her party dined on "part of a raccoon; it was very fat and tasted like lamb if eaten with mint sauce."

If raccoons in the Toronto area were valued for their meat and fur – Davy Crockett's coon-skin cap was a popular item among bushmen – they were also hunted for the sport of it. In a letter from Erindale written in 1832, Thomas William Magrath described "Hunting the Raccoon" to the Reverend Thomas Radcliff of Dublin as "the only hunting of wild animals in which the fair sex partake." When "Cooney" is found in a tree near the house in daylight, he writes, "men, women, children, domestics, dogs, &c." join in the fun:

If there be a gun in question the sport is soon over; if not, the tree must be cut down. Pending the operation, all

eyes are fixed on Cooney, sitting aloft with perfect com-
posure, and looking down with ineffable contempt upon
the gaping enemy; and with some justice! – for how could
he imagine that, with the purpose of destroying a peace-
able and harmless animal like himself, a domestic host
should be arrayed against him? He gives no credit to it,
'till the creaking tree yielding to the axe, begins to give
way, when running rapidly down the stem, and bolting up
that of an adjoining tree, he makes a second effort at
security.

In the confusion upon his first descent, he frequently
escapes; all striking at him together, intercept each other's
implements of war. Cunning and nimble as a fox, he
avoids them all; but should he cling to the falling tree, he
comes to the ground, bruised, and stunned, an easy victim to the beetle, potstick, fleshfork, or poker of the ama-
zonian cook maid, who carries him off in triumph to the
kitchen, encouraged, by her success, to hope for a few
more to line her Sunday cloak with their comfortable
skins.

Mrs. Clever's raccoons, or those in our own back yard,
didn't gaze down at us with "ineffable contempt," but rather
with a kind of bemused interest. Who, after all, was feeding
whom? IQ tests place raccoons slightly behind rhesus mon-
keys in intelligence, which makes them smarter than cats,
dogs, and even foxes. Raccoons can think things out. They
can reason. Rick Rosatte, the wildlife biologist with the Ministry of Natural Resources in charge of the Scarborough
rabies program, says he has watched a trapped raccoon figure
out how to get out of a wire cage fitted with a trap door at one
end that can be unhooked only from the outside. The raccoon reached out through the wire, then back in behind the
trap door, sprung the small release mechanism with its fingers, pushed open the door, and waddled off into the woods.

And there's the story of one female raccoon in North York who figured out how to work an electronic garage door: she pushed the outside button to open the door, then went in and pushed the inside button to close it. And anyone who has tried to keep raccoons out of a compost bin will have stories about failing to do so: a friend of mine fitted the lid of his with a metal clasp that my own seven-year-old daughter couldn't open, and the raccoons were in it in two nights. Another friend outside the city who keeps chickens says that in order to make a chicken coop raccoon-proof, you have to make it human-proof.

Raccoons are no longer thought to be related to bears (as Linnaeus suggested by calling them *Ursus lotor*), nor even to pandas (most modern biologists place pandas in a taxonomic room of their own), but are actually closer to canines, which is why they are now called *Procyon lotor*, *pro* meaning before and *cyon* meaning dog. Their nearest living relatives are the ring-tailed coatis of the southwestern United States, and the South American kinkajou, a beautiful, cinnamon-coloured animal with a long, prehensile tail. Raccoons were part of a vast northern migration that crossed the Panamanian land bridge three to five million years ago, spreading out of South America into Central and North America. Among the twelve other animals native to the southern continent now found here are the opossum, the armadillo, and the porcupine. (The recent increase in opossums in Toronto suggests that the migration is still taking place.)

The bear hypothesis came from the observation that raccoons are flat-footed, or plantigrade walkers, like bears, and their tracks clearly show heel pads more typical of bears than of canines. These tracks are extremely distinctive: look for them in the snow in late spring, or in the soft mud beside the streams in many of Toronto's ravines. The front feet are smaller than the hind feet (unlike canines) and leave star-shaped, human-hand impressions with little dots at the ends of the fingers, made by the tips of their long, curved nails. The prints of the hind feet are longer and deeper (because most of the raccoon's sixteen kilograms are centred over the hind area) than the forefeet, and the toes tend to remain closer together, but they still resemble human footprints.

Raccoons den up in the winter, but they do not go into deep hibernation. They doze during extremely cold weather, then rouse themselves when the temperature rises to go out and forage for food. In the wild, they will seek out hollow trees, shallow caves, old fox dens, and even, in extreme cases, the abandoned nests of large birds; in the city, where hollow trees are not in great abundance and raccoons are, they have to find alternatives and will take over attics, chimneys, garages, garden sheds and culverts, and under porches, piles of debris, and anywhere else that seems safe, warm, and close to food. The females den up with their offspring of the previous year, usually two of them, but the adult males live alone. In February and March, the males make the rounds of female dens within their home ranges and impregnate them one after the other. Males are polygamous, and the females are monogamous; a female, once mated, will reject the advances of all other suitors, thus keeping the male's home range intact. The young are born in early May, sixty-three days after mating (exactly the same gestation period as for coyotes and wolves); the kits are blind and helpless, gaining sight after three weeks and becoming weaned after three and a half months, around mid-July. They stay with their mother until they are about a year old – in other words, until the next gang of siblings comes along – at which time they disperse to set up home ranges of their own. The males take two years to reach sexual maturity; females usually bear their first litter in their first year.

Raccoons in the wild are more or less indifferent to human incursions into their territory; they'll scuttle out of our way, but they don't take off in a blind panic, as prey species such as

deer and grouse do. In the city, raccoons are even more fear-
less – proximity makes the heart grow fonder. We must not,
however, go all anthropomorphic and mistake fearlessness for
friendliness. Raccoons will attack when they feel threatened
or even grouchy. They know that human beings are merely
fellow predators, and since predators rarely eat other predators, raccoons have no particularly compelling reason to
avoid us. But they also have no reason to like us.

We may have a few compelling reasons to avoid them,
however. Raccoons are the bearers of two diseases dangerous
to humans. One is a parasitic roundworm, *Baylisascaris procyo-
nis*, that lives in the intestinal tract of raccoons and the eggs of
which are dispersed in raccoon faeces. In some infected areas,
wildlife rehabilitators are warned to wear face masks when
handling raccoons or raccoon scat, since the eggs can be
inhaled.

The other disease carried by raccoons is rabies.

On July 13, 1819, Charles Lennox, fourth Duke of Richmond
and Lennox and the governor of Canada who preceded Lord
Simcoe, while on a tour of inspection of the country's military
installations, stopped for dinner in a small village north of
York called Mount Pleasant. After dinner, he attended the
raising of the Presbyterian Church, and upon his departure
the next day the village was renamed Richmond Hill in his
honour. From there, the duke travelled Quebec's Eastern
Townships, where another village was renamed Richmond.
While there, he was bitten on the hand by a fox, but the
wound healed, and he continued his tour, travelling up the
Rideau River to Bytown, near which yet another village was
to be christened Richmond. On his way to this Richmond,
the duke began to complain of pains in his neck and shoulders. At dinner, those in his party noticed that he ate nothing,
"but played in his fingers a little bullet of soft bread." At one
point, he emitted an involuntary scream, "like the snarl of a

dog," after which he excused himself, muttering "Shame, Charles Lennox, for shame Richmond, remember!"

After a sleepless night, the duke continued towards Richmond, but was soon so weak that his party was obliged to stop at the hut of a Sergeant Vaughan, at the edge of a large swamp. The next day, August 26, they rode on to Richmond, and although the duke was able to walk about the village, he was obviously in great distress. They left early the next morning, and after walking six miles (the duke was now too weak to ride a horse) through a heavily wooded area they came to a spot where the road crossed a small stream. By this time the duke was being supported under each arm by Colonel Francis Cockburn and a second man named Bowles, who described him as being "extremely feeble." At the sight of the stream, however, the duke threw both men from him "with prodigious strength," jumped over a six-foot log fence beside the road, and ran off into the woods, hotly pursued by his two companions. Lord Dalhousie, who heard the story directly from Colonel Cockburn and recorded it in his diary, continues:

> They followed him, but getting into deep boggy undrained land, full of holes & large burnt stumps of trees, they had the utmost difficulty to keep up with him. The Duke rushed at these & over the fences with furious violence, at last he reached a Barn which was open, & perhaps about 300 yards from where he broke away; he threw himself down on some straw, as if quite exhausted by his exertions; still perfectly sensible, he refused obstinately to go into the hut, because it was a few paces nearer the river. Nature however could do no more; in the Evening he became drowsy & insensible to anything; they carried him into the hut & there he died next morning, on 28th August 1819.

Lennox was the first recorded rabies victim in Canada, and it is fitting that he died in Ontario from a bite that he received

in Quebec. Until recently, southern Ontario was known as
the rabies capital of North America. More cases of rabies per
capita have been reported in this 100,000-square-kilometre
strip – an average of 3,000 people in Ontario are treated for
contact with rabies each year, although the number of deaths
approaches zero – than in any other area of comparable size
on the continent. Since rabies from Arctic foxes filtered down
from northern Quebec in 1948, there have been more than
50,000 cases of rabies in southern Ontario, about 85 percent
of all the cases in Canada. In 1986, Rick Rosatte began an
intensive campaign to inoculate foxes in eastern Ontario. By
wrapping thousands of vials of rabies vaccine in hamburger
patties and dropping them from aircraft, he effectively eliminated fox rabies in that region.

Ontario is no longer the top hot spot for rabies, not
because the MNR has controlled fox rabies, but because raccoon rabies has boosted the figures elsewhere. Last year, of
New York State's 2,747 cases of rabies in animals, 2,369 of
them were raccoons, and "most of the rest," says Charles Tri-
marchi of the New York Department of Health, "were other
animals that had raccoon rabies," including eighteen white-
tailed deer. More than 3,000 humans were treated for rabies,
almost three times the number treated in 1992 and thirteen
times the number treated in 1990. When raccoon rabies
reaches southern Ontario, we could easily regain the lead.

"That's Barnie," says Mrs. Clever, handing me a Polaroid of a
fat old raccoon squatting on her driveway. "And this is Tickle-
tummy. He was a bold one, he was. He used to come right up to
the patio whenever we had guests over for cocktails, and he'd
climb right up into people's laps to get at the hors d'oeuvre."

"And this is Barnie II," she continues, handing me another
photo. This one shows a portly, aged raccoon at the base of a
maple tree. His tail is clearly missing. "He's about eight years
old in that picture. He hasn't come back this year, so I suppose

he died over the winter. They don't live much past ten. He came out in February, during that vicious cold snap we had, and I suppose he shouldn't have. He used to come first, before all the rest, and I'd give him bread dipped in honey as a treat, and then he'd leave and come back with the others for the kibble. Then when his teeth went he couldn't eat the kibble. I last saw him in the middle of March, and he had lost his ears, and his tail, and his teeth. I guess by now he's in Raccoon Heaven."

According to a report published in 1992 by Ontario's Raccoon Rabies Task Force – a panel of experts, including Rick Rosatte, from several federal and provincial ministries set up to develop strategies for dealing with the expected invasion – in the fall of 1991 raccoon rabies was only eighty kilometres south of Buffalo, New York, just across the Niagara River from Canada. Unless something is done to stop it, raccoon rabies in Buffalo means raccoon rabies in Ontario. "We may expect this strain to arrive along the New York side of the Niagara River in late 1993 or 1994," the Task Force predicted, accurately as it turns out: by March 1994, reports of raccoon rabies were coming in from just twenty kilometres south of Buffalo. "Although the Niagara River may seem a barrier to crossing by infected raccoons," continued the report, "raccoon rabies has crossed the Susquehanna and Hudson rivers at points where they are as large as the Niagara." Raccoons are very comfortable with water, can walk across ice, and are quite capable of hitching rides in buses, boats, and camper vans. Last year, a raccoon was caught hopping out of a transport trailer at the Ontario Food Terminal; it was the merest fluke that an Animal Control officer happened to spot it. "Therefore," the report concluded, "Ontario should be prepared for the imminent arrival of a rabies strain spread by raccoons."

But how do you prepare for an epidemic for which there is no effective antidote, and one that is carried by an animal almost as ubiquitous as the house cat? That was Rick Rosatte's problem. Raccoon density in Scarborough averaged about twelve per square kilometre, and when Rosatte checked the cell that included the Toronto Hunt Club, he turned up sixty-six raccoons in one week. The club manager's wife, it turned out, was feeding thirty of them every night, letting them walk right into her living room. Mrs. Clever has worked closely with Rick as well as with the Toronto Humane Society (THS) over the years, but some people become very defensive about their attachment to raccoons, regarding wildlife officers as enemy agents out to destroy their pets. Two years ago, in an effort to curb the spread of raccoon rabies in New York, the state's Department of Environmental Conservation told wildlife rehabilitators – who are licensed to treat sick or injured wild animals – that they had to report any raccoons brought to them by the public. Thinking that the department intended to "euthanize" the animals, Barry Rothfuss, a wildlife rehabilitator who had thirty-seven raccoons living with him in Peekskill, in the southern tip of the state, packed them into a yellow Ryder van and headed north to release them in Tioga County, where raccoon rabies had not been reported. He'd released half of them by the time the authorities caught up with him: "It's a little depressing," he told reporters, "because you never know what their fate is once they're gone, and you do get attached to them." The next year there were fifteen cases of raccoon rabies in Tioga County.

Nathalie Karvonen, who operates the Toronto Wildlife Centre, a privately run clinic that provides medical treatment for wild animals, is worried that a similar dictate from the MNR in Ontario would have equally harmful results. "People won't bring in sick animals if they think we're going to kill them," she says. "Just last week a woman called me, she wouldn't give her name or her address, and said she had a sick raccoon and wanted me to tell her what to do with it. I told her to bring it in here, and she said no, she wouldn't do that. She said the raccoon had already bitten her four times on the hand, drawing blood each time, and she was afraid I'd test it for rabies, which meant I'd have to euthanize it – the only way to test for rabies is to take a sample of the brain tissue. She just wanted me to tell her how to treat the raccoon. I told her she needed to be given immediate post-exposure treatment, but she just said, 'No, I'll be fine, I'll be fine,' and hung up. It's very disturbing."

But the danger of preparing two million people for a new wave of rabies is that individuals will become so afraid of raccoons that they'll take matters into their own hands. In April 1994, a Toronto man was fined $100 for causing undue pain and suffering to a raccoon – he had built himself a wooden trap armed with 300 seven-centimetre nails and weighted down with rocks and a four-and-a-half-kilogram sledge hammer, and put it in his back yard to keep raccoons away from his grapes. Wendy Hunter at the THS Wildlife Department said the contraption looked like "a medieval torture machine" when it was brought in to her. The raccoon caught in it had to be destroyed. And there are reports from the United States of vigilante groups threatening to cruise the suburbs and shoot raccoons on sight. The very friendliness of raccoons could tell against them, as one of the first signs of rabies is an apparent loss of fear of human beings.

Rick decided to prepare Ontario by devising a way to stop raccoon rabies at the Niagara River. His plan, which initially was intended to go into effect in July 1994, would cost the province about $150,000. The money, however, had to come from the provincial government, and the provincial government didn't have any money. Rick and the other members of the Task Force applied for it anyway and then set about trying to find a vaccine that would work on raccoon rabies.

Rick's rabies research lab is located in an MNR complex on the Rouge River, just outside Maple. It's a large, bare room in Building 32, ringed with cupboards and counters on which a few empty cages sit below wall maps of Ontario and upper

New York State. Rosatte's office, built into the corner, is cosy enough that a map of Toronto takes up nearly half a wall. On it, each borough is shaded a different colour and divided into square-kilometre cells: Scarborough, for example, comprises sixty of these cells.

"There are four technicians working this whole area," says Rosatte, indicating a chunk of Toronto that extends roughly from the 401 down to the lakeshore, and from Victoria Park Avenue over to Markham Road. "It's called the trap-vaccinate-release, or TVR program. We go into a cell with seventy-five live-traps, and we leave them there for a week, checking them each day and vaccinating and ear-tagging every raccoon we catch. Then we release them. The following week, we place twenty-five more traps and see how many tagged and untagged raccoons we catch. We're pretty sure we're getting 60 to 70 percent of the animals. We haven't had a single case of skunk rabies in that block for five years now, and it still persists out here," he says, pointing to the rest of Toronto, "where we're not vaccinating. So we know the system works."

The vaccine with which Rick and his technicians inoculated Scarborough's raccoons is called Imrab, the commercially available rabies vaccine that is also effective on domestic dogs, cats, and cattle. Imrab isn't licensed for use on wild animals — no rabies vaccine is licensed for use on wild animals in North America — and every time he uses it he has to get permission from Agriculture Canada. It means a lot of red tape, but if it is injected into 60 percent of the raccoon population, it will stop the spread of rabies, at least in an area as confined as Scarborough. But TVR is an impractical program for dealing with a general epizootic of raccoon rabies covering the 100,000 square kilometres that make up the Niagara Frontier. If Rick's plan was to work, he'd have to find some other method of getting rabies vaccine into raccoons.

At first, the answer seemed to be an oral vaccine called ERA, developed by Connaught Laboratories in North York in

1988 specifically for Rick's fox program. Since 1991 it has been manufactured by the Langford Laboratories in Guelph. It is a modified live virus, very similar to the vaccine used to inoculate humans in that it simulates rabies in the recipient, fooling the immune system into producing an army of antibodies to ward off what it thinks is an invasion of rabies virus. Like the rabies virus itself, ERA works through the salivary glands, only in the opposite direction. It has to be taken into the animal's mouth and spread around so that it gets to the glands before it is swallowed; otherwise, the acids in the animal's stomach would kill the ERA.

Rick's first problem was finding a bait into which he could insert capsules of ERA. The bait he used for foxes didn't seem to appeal to raccoons. "It's made of beef tallow," he says, "with some kind of wax in it to hold it together, and vegetable oil and chicken essence mixed in for flavour. Foxes love it. They take it up in their mouths, chew on it for a while, their teeth puncture the capsule, and the vaccine spreads around in their mouths." He shows me a capsule that has been chewed by a fox: it looks pretty mangled. "Raccoons, however, seem to be able to take it or leave it, mostly leave it. So we changed the formula." He takes a small cube from the drawer of his filing cabinet and hands it to me. It is soft and cream-coloured, like white chocolate. It smells like almond bark. "We make them with marshmallow and icing sugar instead of chicken," he says. "We developed them here in the lab; I tested them on Mrs. Clever's raccoons last fall, and they seem to work just fine. We had very good acceptance with them – around 58 to 60 percent."

His second problem, however, was that the ERA vaccine doesn't work on raccoons. "There seems to be something about a raccoon's immune system that's different from that of a fox," he says. "If you take ten foxes and give them 1 cc of the ERA vaccine, nine or ten will be protected; if you give it to ten raccoons, only two or three will be protected. So we had a bait that works; we just had to find a different vaccine to put in it."

While thinking through the vaccine problem, Rick also began pestering various provincial ministries for money. Even if he found the right antidote, he says, he wasn't hopeful of getting the go-ahead. "What we proposed to the ministry was to vaccinate this whole block right here," he says, running his hand along the strip of land between the Niagara River and the Welland Canal – the Niagara Frontier. "We can run a TVR program in there fairly easily, capture 60 to 70 percent of the raccoons, inject them with Imrab and release them. Then we'd do a drop-bait program all over here." His hand moves from the Welland Canal, around the end of Lake Ontario, to Hamilton. "With raccoon rabies only twenty miles from the border, we didn't feel we had a lot of time. If we had ten trappers, we said, we could do it in two months. If we only got enough funding for five trappers, it'd take four months. But it'd only cost the government about $150,000. If we don't do it this summer, and raccoon rabies crosses into Ontario, it'll cost us millions, and the number of cases of human exposure to rabies we'd have to treat would double, maybe triple."

It was the drop-bait part of the plan that was problematic. While waiting to hear from Queen's Park, however, Rick learned of a new vaccine that would work on raccoons. The vaccine, however, was the subject of so much controversy that he feared he might never be allowed to use it. Called "vaccinia rabies glycoprotein recombinant vaccine," or VR-G for short, it had been tested in Europe and the United States and found to be just about 100 percent effective; in fact, the Wistar Institute of Anatomy and Biology in Philadelphia was developing a similar product for humans. However, VR-G has not been licensed for use on any animal yet, and some health officials believe it shouldn't be.

VR-G is a form of recombinant DNA and as such represents the potential for a new life form. Outside of science fiction, scientists are generally hesitant about releasing new life forms into the environment. New viruses keep immunologists lying awake at night in a cold sweat. As Rick puts it, "Let's say there's a raccoon out there with some kind of pox virus, let's say raccoon pox, and you give it a vaccine with some of this recombinant DNA in it, and the vaccine joins up with the raccoon pox to form a whole new organism that we don't know anything about, and who knows what the hell will happen?" One thing that could happen is that this new organism would be a virus worse than smallpox and for which there is no known antidote. In other words, we could end up, once again, with a cure for rabies that is worse than the disease.

"Even so," says Rick, "all the tests showed nothing." The controversial vaccine had been effective against fox rabies in Belgium and Luxembourg, and against raccoon rabies in Virginia and South Carolina. No new diseases had sprung up anywhere. The vaccine was being manufactured in France and Georgia, and McMaster University was thinking about getting into the act as soon as licensing went through. "The lab and field work had been done," Rick adds, "and it was just up to the politicians. If we could get it, we'd use it."

The announcement came in early May 1994: the MNR appropriated $500,000 to Rosatte and the Rabies Task Force to try to stop raccoon rabies from crossing the Niagara Frontier. The minister of natural resources even travelled to Niagara Falls himself to make the announcement: Rosatte, who also made the trip, says that he got even more than he had applied for, and he now had enough money to conduct a TVR program in the Frontier as well as more research into the VR-G vaccine at Agriculture Canada's research laboratory in Nepean.

As soon as I heard the news I phoned Rosatte, who sounded jubilant and busy but not exactly victorious. He had placed two trappers in the Frontier on May 2, each with 125 traps, which they were checking every day. They vaccinated the raccoons, placed ear-tags on them so they could keep track of which ones were done and which weren't, and then released them. On May 30, Rosatte sent three more trappers into the area. "We'll have the whole area covered by September 21," he said. And if the Nepean research leads to the licensing of VR-G, he intends to conduct a drop-bait program all the way from the Welland Canal to Grimsby. He also said that the City of Toronto Health Department had applied to the provincial government for additional funding to conduct a TVR program in the city, and that Metropolitan Toronto had done likewise.

"So you got everything you asked for," I said to him. I was trying to account for the tone of caution I could detect in his voice.

"Yeah," he said, "it looks pretty good."

"And you are confident you can keep raccoon rabies out of Ontario?"

"No way," he said quickly. "This is just a kind of containment strategy. There's no way we're going to keep it out."

The same week that the environment minister made his announcement in Niagara Falls, Rick said, a tractor trailer with a load from New Jersey pulled into Brampton, and when the driver opened the rear door out hopped a raccoon. It was caught, tested, and did not have rabies, "but the writing is on the wall," Rick added. "Nothing we can do will keep that first case out of Ontario. It will get in."

Mrs. Clever tosses a final scoopful of kibble into the darkness and closes the sliding door. "There," she says, "that's enough for tonight."

"If raccoon rabies crosses the border," I ask her, "Will you still feed your raccoons?"

Mrs. Clever is not one to close her eyes to a problem. A few years ago, when one of her raccoons showed up one evening in an obviously diseased state – it was wobbling on its feet and could hardly pick up a piece of kibble – she got it into a cage and called Paloma Plant at the Toronto Humane Society. "Paloma came and got it and euthanized it and shipped it off to Guelph for analysis," she says. "We waited a full year and didn't hear anything, Guelph is always so very slow, so finally I phoned the university myself and asked to speak to the vet, and he told me the poor thing had been riddled with roundworm, the worst case he'd ever seen. They'd gone right up into its brain. Paloma gave me some medicine after that, to give to the others, so every night I gave them two drops of medicine on little cubes of brown bread before they got their kibble, and we've never had another case of roundworm since."

But this time she looks anxiously towards the glass door, where nine raccoons are crunching kibble in the dark. "It won't come to Toronto," she says simply.

I could tell she was uneasy with my question, and I thought I knew why. It was the wrong question. It showed her that I hadn't really understood about her and the raccoons. Either we are part of nature or we're not, she seemed to be telling me, and if we're part of it, it isn't just when it's safe, or convenient, or pretty. Nature isn't something we can opt out of when it suits us. Mrs. Clever made me understand that, and I was grateful to her for it. I thought about the raccoons in our own back yard, and I hoped they would come back.

THE SPARROW'S
RISE AND FALL

O n a sunny day in early March, while I was standing on
a corner of Bloor Street waiting for a light to change,
I watched a pair of house sparrows doing their annual spring cleaning. House sparrows are cavity dwellers –
they like to build nests in boxes or holes rather than jam them
into the forks of branches – and these two had found a mag-
nificent cavity in the wall of a grocery store, just where the
hose from a roof-top refrigerator condenser disappeared into
the side of the building. Above the hose, an opening about the
size of a credit card gave the birds access to the wall's interior,
and to judge from the droppings down the side of the building
and along the hose, and from the amount of nesting material
sticking out of the hole, the cavity had been well used, quite
possibly by the same pair of birds, for several years.

Re-using old nests is a problem for songbirds, mainly
because of what birders euphemistically refer to as "nest
fauna" or ectoparasites – insects that inhabit birds' nests in the
hope of eventually inhabiting the birds themselves. Nests
attract insects because they provide warmth, concealment,
and nutrition in the form of blood, guano, and food droppings. Literally thousands of species of flies, fleas, lice, bedbugs, assassin bugs, ticks, and mites lay their eggs in birds'

nests, and a single nest might contain two dozen species of insects. For example, some ticks (which are a subgroup of mites and thus not really insects, but spiders) fall off their hosts after each blood meal and have to find a new host every time they get hungry. What better way than to hang around in a host's nest and hop back on again whenever the host returns to sit on its eggs or feed its nestlings? You can even snack on the nestlings while she's gone. Several species of flies lay their eggs in birds' nests; the eggs hatch, become larvae (maggots), and progress to the pupal stage within the breeding season of the host bird, then leave the nests as adults at about the same time the nestlings do. One louse fly (family Hippoboscidae, of which there are more than 100 species) over-winters in nests as pupae and hatches when the occupants return next spring to re-nest, which is why re-nesting is a problem for birds: the adult hippoboscid spends its entire life on the host bird. Adrian Forsyth, in his book *The Nature of Birds*, describes the harrowing experience of removing a multi-year eastern phoebe nest from his woodshed one summer, after the fledglings had flown: the nest in his hands was little more than a huge ball of writhing, purple-grey lice.

It would therefore seem to be to a bird's advantage to re-settle each year in an entirely new nest, but it's not as simple as that. Building a new nest costs a lot of energy, especially for birds that return in late February and early March, when the temperature can drop at night to below freezing (the night after I observed the nest-cleaning house sparrows, the temperature went down to -18°C). Birds need to store up energy during such sunny days, not squander it in a search for new nest sites and material. For city birds such as house sparrows, the number of good nesting cavities is not as high as one might think. Older houses provided lots of holes and crannies for nests, but modern buildings, with their emphasis on air-tight construction and space-age materials, don't. Competition for prime real estate is strong in the bird world, and the two house sparrows I watched at the grocery store were defending their spot like a pair of Spartans at Thermopylae: either the male or the female remained on guard outside the hole while the other went inside, and whenever another bird alighted nearby the defender would puff up his or her chest feathers impressively and scold the intruder away.

Rather than go to the trouble of building new nests, then, some birds have developed strategies for dealing with parasitized nests. Starlings, for example, add new green vegetation to their old nests each year, apparently selecting specific aromatic plant species, such as yarrow, that have a tendency to repel or even destroy nest parasites. The Latin name for yarrow is *Achillea millefolium*, or thousand-leaved Achilles: Greek soldiers (and later, those of the American Civil War) used to pound its leaves into a paste and apply it to wounds to stop bleeding; like most of the Asteraceae, however, including ragweed, yarrow also has insecticidal properties. House sparrows will add plant material to their nests, too, although they seem to prefer wild mint and mustard. There is even the extreme case of the eastern screech-owl, which actually carries a species of snake, known as the blind snake, to its nest to feed on blow-fly maggots.

These strategies are all right for summer nests, but what about nests built in early spring, when herbs, aromatic or otherwise, are not available? My house sparrows seem to have solved the problem by simply cleaning out some of the old nesting material. While one of the pair, usually the female, stood sentry duty outside, the male would hop into the cavity and periodically hop out onto the hose perch with a single piece of dried grass or twig in his beak. After looking nervously around for a few seconds, he would drop his load and then disappear inside for another beakful. The snow at the base of the wall was littered with nest debris and, I'm willing to bet, the unhatched eggs of some over-wintering species of nest fauna. Since house sparrows don't usually lay their eggs until April, there was still plenty of time to rebuild the nest with fresh material before the serious business of egg-laying and chick-rearing began.

The house sparrow (*Passer domesticus*) is a small, mostly dun-coloured songbird usually lumped by birders into what Tom Hinks, chief ornithologist at Point Pelee, calls the "LBJ" category – "little brown jobs." At Point Pelee, or in Toronto during the spring migration, when more flamboyant species of birds are around, seeing a house sparrow may not be all that exciting. But in winter and especially early spring, when they have the city pretty much to themselves and the leaves are not yet hiding them from view, they are so numerous and visible that watching them can afford hours of entertainment. True, watching house sparrows in the absence of yellow warblers might be a bit like tuning in to in-line skating hockey during an NHL players' strike, but the contemplation of any species of wildlife has its rewards, for the soul as well as for the eye. Walk past a thick hedge or evergreen bush in winter, especially one that is in direct sunlight and close to a brick wall, and you will suddenly hear the joined chorus of dozens of house sparrows animated by the borrowed solar warmth, hopping from twig to twig, chattering away like kids at summer camp, safe inside their sanctum: it is a sound that lifts the spirit.

House sparrows are easy to identify. About 120 mm long from tail to the tip of their strong, seed-cracking beaks, they are brown on top and beige underneath, with a single white wing stripe on the upper coverts. The males have white cheeks with grey patches under the eyes, a grey cap running right back over the crown, and a large, black bib extending from their beaks, down their throats and spreading out over their chests. Look for this bib – it is the house sparrow's most distinguishing feature. The females don't have it – they are of a more uniformly drab brown, with a lighter brown streak running directly back from the eye – but wherever you see a female you'll usually spot a male close by.

House sparrows are highly gregarious, gathering in huge flocks and building nests in close proximity to one another. They make their nests from February to May in any convenient cavity with an opening of at least 5 cm – of 1,200 nests examined in one study, 529 were in bird boxes (many of them designed for some other bird, such as purple martins or eastern bluebirds); 420 were in holes in buildings; and the rest were found in natural tree cavities – formed when a large branch breaks off and its stub rots away into the tree's trunk – or in woodpecker holes in trees, fence posts, and hydro poles. But they can be found virtually anywhere: in porch corners, or in the gingerbread trim in the eaves of older houses, in the ivy on house walls, on light fixtures, beams, ladders. The nests themselves are very loose, squashed-ball-shaped affairs about 35 cm in diameter and are composed of tough building materials on the outside – grass, string, bark, stalks, paper, plastic, cloth, roots, leaves – and a relatively soft lining of feathers, usually, but also plant down, hair, paper, cloth, and moss.

There are thirty-six species of true sparrows in North America, but the house sparrow isn't one of them. What we refer to as house or English sparrows are in fact neither English nor true sparrows. For a while they were considered members of the Old World weaver finch family, the Ploceidae, of which there are 263 species worldwide, most of them in Europe and Africa. But a few years ago they were shifted into the Passeridae family, along with another introduced species, the Eurasian tree sparrow (*Passer montanus*). The introduction of the house sparrow to North America is without doubt the most successful transplantation of a new species ever conducted, on a par with the release of rabbits in Australia. And for a while, at least, it was attended with equally disastrous results.

In the fall of 1850, the directors of New York's Brooklyn Institute (now the Brooklyn Museum, a natural history academy) imported eight pairs of English sparrows, as they were called (because they were imported from England). The birds were kept in cages at the institute over the winter and were released in New York City in the spring, when they promptly died. The next year, Nicolas Pike, one of the institute's directors, stopped off in England en route to Portugal and ordered "a large lot of Sparrows and song birds" from Liverpool to be shipped back to New York aboard the steamship *Europa*. Fifty of this lot were released from the ship as soon as it entered the Narrows, and the rest were placed in the chapel tower in New York's Green-Wood Cemetery when the ship docked. No one knows what happened to the first fifty, but the enchapelled birds, like their compatriots of the year before, did not thrive, and were taken back indoors that winter by John Hooper, the cemetery's groundskeeper, and were released again in the spring. This time, according to Pike's report, "they did well and multiplied."

That has turned out to be something of an understatement. In fact, the house sparrow spread across North America like a prairie grass fire. Within a few years they were turning up whole states away from New York. There are several predictable reasons for such rapid expansion, although no one predicted it at the time. House sparrows are prolific and tireless egg-layers, laying up to seven eggs at a time and often producing a second clutch before the fledglings from the first have left the nest. They seem to have no climatic restrictions, flourishing as far south as Texas and as far north as Kapuskasing with only minor changes in feeding habits. And, initially at least, they were mollied and coddled by earnest but ill-informed amateur birders who, like Pike, thought they were doing something environmentally correct by introducing the bird on this side of the Atlantic. They were wrong.

In 1854, a third batch of house sparrows was imported and released in Portland, Maine, by Colonel William Rhodes of

Quebec City, who was the Commissaire de l'Agriculture for what at the time was still known as Lower Canada; why he was meddling with the biodiversity of Maine is anyone's guess, but there he was. He wasn't alone. In 1860, six pairs were "liberated" in Madison Square, New York City; in 1866, 200 more birds were "acclimatized" in Union Park, also in New York City. In 1868, fifty were set free in Canada – in Quebec City's Jardin du gouverneur, by the same Colonel Rhodes. By 1876, house sparrows had been either imported and released, or else had spread by their own inclination and ingenuity into thirty-two states and eight Canadian cities, including Toronto.

The first house sparrows to appear in Toronto probably came from a colony that had been established in Strathroy, a small town just west of London, in March 1874. "I sent to a New York bird dealer," recalled amateur ornithologist L.H. Smith of Strathroy in 1888, "and he forwarded me per express, twelve birds, six males and six females, at a cost to me of $1 each. If all the Sparrows in our town are mine, and my neighbours all say they are, then I have at least plenty for my money." Indeed he had, for by 1886 there were thousands of house sparrows not only in Strathroy, Smith noted proudly, but also "in every town, city, and village in this part of Ontario. . . . It was only a few years after 1874 that I noticed them at Toronto and London and other places east of this, and I have no proof that all did not come from my six pairs."

This madness to assimilate the house sparrow was under-taken so assiduously partly because house sparrows were thought to be pretty birds with pleasant songs; native North American sparrows had not yet adapted to city life, and Pike and his followers believed house sparrows would add a touch of wild and harmonious beauty to our increasingly charmless urban cores. They may have been right about the pretty part – house sparrows, as I implied, are an acquired taste, perhaps best relished in the absence of the showier species. But euphonious songbirds they ain't. Possibly because it was known (rightly) that native sparrows, like the white-throated sparrow and the song sparrow, are accomplished singers, and it was assumed (wrongly) that the house sparrow was in fact a sparrow, people thought house sparrows were likewise proficient in the performing arts category. Not, alas, so. House sparrows have nice voices, but they choose to exercise them in relatively mundane fashion. Passing a bevy of chirping house sparrows on a cold, grey winter day can pause the spirit momentarily in its downward spiral towards Absolute Zero, but that is mainly by default. The true singers have quite sensibly taken their throats elsewhere for the winter. Song is to the English sparrow what vegetables are to English cooking.

The second reason for the fervour of house sparrow importation was equally wrong-headed. House sparrows, for reasons that even at the time were unclear, were thought to be voracious insectivores that would rid our cities of leaf-eating caterpillars – particularly those of such geometer moths as the spring cankerworm (*Peleacrita vernata*) and the elm spanworm (*Ennomos sussignarius*) – which were in turn thought to be devouring urban shade trees at an alarming clip. Their reputed insatiable appetite for insects was also supposed to be a boon to farmers. In 1868, a few days after Colonel Rhodes released his flock of house sparrows in Quebec City, the *Courrier du Canada* commended the good colonel "for the initiative he has taken to augment the number of our birds that are useful to agriculture." Why anyone thought house sparrows would eat masses of insects is a mystery, since they were well known as grain-robbing pests in Europe. But natural history was a young science, and birds were just generally considered to be higher up the food chain than caterpillars. Perhaps female house sparrows had been observed feeding insects to their young in the spring: all passerines stuff bugs down their young's throats for the first week or so, until the chicks are strong enough to survive on predigested seeds regurgitated by the parents. But at any other time of year, a house sparrow will choose an oat over an inchworm.

Or perhaps the importers had been influenced by litera-ture. In 1868, a Belgian poet named Jean Lesquillon published a poem entitled "La Proscription des moineaux," roughly, "The Banning of the English Sparrows." The 420 lines of the poem recount the illuminating story of Belgium's long war of attrition against grain-eating songbirds, the English sparrow among them. In the poem, the Belgian government issued an edict requiring all citizens to use any means to dispose of as many English sparrows as possible. As a consequence, English sparrows soon disappeared from the Belgian countryside. Soon afterwards, however, Belgian farmers were plagued by hordes of insects, which proliferated stupendously in the absence of their natural foes. Farmers beat their chests and tore out their hair as they watched their figs and pears, their strawberries and mulberries, being slowly devoured by ram-paging caterpillars and weevils. Then, just as the situation seemed hopeless, a lone English sparrow fluttered into view; the farmers ran after it, begging it to go get the rest of its fam-ily and return to Belgium. The brave little sparrow did as he was bid, and all was well. It was a fine poem. It was published in France in the very year Colonel Rhodes, Lower Canada's director of agriculture, ordered fifty English sparrows for his city – from France.

At any rate, the mistake soon became apparent as millions of house sparrows invaded North American cities, towns, and farming districts, destroying gardens, orchards, flower beds, and grain fields. In 1888, the same year Mr. Smith of Strathroy, Ontario, was patting himself on the back for being responsible for Toronto's house sparrow population, the U.S. Department of Agriculture's Division of Economic Ornithol-ogy and Mammalogy issued its very first publication: it was called *The English Sparrow in North America.* "There can be no question," it intoned, "that a thousand times as much energy and money have been spent already in fighting Sparrows in America as were expended in introducing and caring for them at first." Farmers in Great Britain, it went on, were reported to have "stared in blank amazement" upon hearing that North Americans had introduced house sparrows and "failed to account for the action except on the assumption that America had been visited by a wave of temporary insanity."

There followed a heated battle between house-sparrow apologists and house sparrow mud-slingers. In an article that appeared in *Field and Stream* on April 5, 1888 – a busy year for house sparrows – Ernest E. Thompson (who later changed his name to Ernest Thompson Seton) wrote that "it has often been argued that, so far as we Canadians are concerned, the Sparrow can never give us much trouble, as the climatic and other conditions are sufficient to prevent its increasing to the same extent as in England. But unfortunately the facts are sufficient to entirely dispel this illusion." Seton, who went on to become the provincial biologist for Manitoba and, later, one of the most renowned nature writers in the world, can be trusted here. He reported seeing his first pair of house sparrows in Toronto in 1874, and in the intervening fourteen years had watched the species increase until it had invaded every town in southern Ontario and was about to "spread over the intervening farm lands" as it had earlier done in England.

But, he continued, was this necessarily a bad thing? Did the house sparrow do as much damage to farmland as it was being accused of by the United States Department of Agriculture? One of the real impediments to determining this was an almost total lack of scientific information about what the house sparrow actually ate, and when. Most of the USDA's data came from farmers, who wanted to be recompensed for their losses and might therefore be just a tad eager to assign damages to the house sparrow rather than to, say, acts of God. To offset the imbalance, Seton solicited the aid of his old friend and mentor, the great entomologist William Brodie.

Brodie was himself a transplant to the New World: he had
been born in Scotland in 1831 and had come to Canada at the
age of four. He was a dentist by profession, in fact he was the
first dentist in Toronto to use chloroform, a kindness that no
doubt endeared him to the general populace more than his
contributions to their knowledge of tree galls, his other specialty. In 1877, he had helped found the Toronto Entomologi-
cal Society, and in 1888 he was an internationally recognized
authority on plants and insects. Seton had been his friend for
a number of years; two years after this article appeared, Seton
was travelling in Manitoba with Brodie's son Sam when the
latter was drowned trying to cross the Assiniboine River. The
important thing about William Brodie here is that he was a
lifelong champion of biological (as opposed to chemical) controls on insect pests. If anyone was going to go out of his way
to praise house sparrows, if there was a chance that house
sparrows might be significant predators of agricultural pests,
it was Brodie.

Seton and Brodie examined the stomach contents of 120
house sparrows shot, mostly by Brodie, in various parts of
Toronto – Brodie's diary for the summer of 1887 lists "3 birds
shot in St. Matthews ward," "5 birds shot on the Don Flats,"
"East Toronto," "Rosedale," "15 birds killed with two dis-
charges of the gun at Greenwood's Crossing," and so on,
throughout the city and its environs. The discharging of shotguns within city limits was a much more common occurrence
then than it is today, it seems. Seton's contribution was mathe-
matical. He gave each stomach a total of ten points, depending on whether he thought the contents were beneficial or
harmful to humans – in other words, a stomachful of insects or
weed seeds got the bird ten points for; a stomachful of wheat
or flower buds got it ten points against. Mixed contents got
mixed points – 60 percent oats and 40 percent grasshoppers,
for example, gave the bird six points against and four for.
They found that 39 percent of the sparrows had some insect

material in their stomachs, and adding up the total points they awarded the bird 795 points for and only 534 points against. As far as Seton and Brodie were concerned, *Passer domesticus* was a beneficial bird. And the case was closed.

Not quite. Seton and Brodie's findings were put down by the bird's detractors as, at best, a case of regional disparity, for studies carried out in the United States came to a quite different conclusion. Examination of the stomachs of 4,848 birds taken from 1879 to 1925 showed that 60 percent of the house sparrow's diet consisted of livestock feed — wheat, oats, etc., taken from cattle feedlots; 18 percent was cereals, or grain taken directly from the field; 17 percent was weed and grass seeds, mostly ragweed, crabgrass, bristlegrass, and knotweed; and only 4 percent was insects. That made it 78 against and 22 for. Subsequent studies have tended to support this more damning conclusion.

More disturbing to bird lovers than the house sparrow's appetite, however, was its invasive behaviour towards native songbirds. So well did the sparrow take over nesting cavities, and so strenuously did it defend them from other birds, that native species declined almost as rapidly as the sparrows proliferated. The most affected native species was the eastern bluebird. Once as numerous as the sparrow — and ironically a highly voracious devourer of caterpillars — the bluebird is now listed as rare in Ontario. And it is far from alone: a survey conducted in (guess when?) 1888 showed no fewer than ninety-four species of native birds that were being "molested" by the house sparrow. In addition to the bluebird, the list included robins, house wrens, purple martins, swallows, chipping and song sparrows, orioles, bobolinks, hummingbirds, and even woodpeckers, although how a house sparrow could molest a woodpecker is beyond me. In Toronto, Seton said he had seen pewees, chipping sparrows, white-breasted swallows, house wrens, robins, and bluebirds "dispossessed by the invader." Clearly, whether house sparrows dined on ragweed or No. 1

Marquis wheat, Canadian ornithologist Percy A. Taverner was speaking for the entire ornithological community when he dryly remarked in his *Birds of Eastern Canada* (1922) that the introduction of the house sparrow in 1854 was "without doubt a mistake." Taverner went uncharacteristically further in his condemnation: "Constant endeavour will keep the numbers reduced but only continent-wide persistent effort will destroy them altogether. Traps, poison, and systematic destruction of the nests are the most satisfactory means of control."

In Montreal, a certain Father Pierre Fontanel, S.J., agreed wholeheartedly with Taverner's assessment. In fact, Father Fontanel had already devoted ten years of his life to eradicating English sparrows from around his residence. From 1919 to 1924, he published no fewer than ten long articles in the *Naturaliste Canadien*, each one delineating in fine detail the escalation of his own war against the English sparrow. "If we don't destroy these little cannibals," he warned, "we will be eaten alive by them ourselves!" The good Father was possessed of a fanatical zeal: "In order to annihilate them," he wrote, "we first have to get to know their habits, their customs; that way, we will be better able to take them by surprise and kill them more effectively. . . . I myself have based my plans for their destruction upon my profound knowledge of their ways. I have spent the past three years perfecting my plan, trying it out, modifying it, simplifying or complicating it as the need arose. But throughout these three years one thing has continued to confound me: the English sparrow has not gone away. The English sparrow does not die."

Father Fontanel called on all the good citizens of Montreal to join him on his crusade. The use of rifles ("both of powder and air"), poison, traps, even bare hands was encouraged: "Hands can be employed to destroy nests, to entrap nestlings, to capture adults when they're sleeping and to wring their necks." Destroying nests was a form of psychological warfare, like aerial bombing in World War II: "Destroying nests negates all the energy they spent making them, and ruins their hopes for the future." Traps, he felt, were the least satisfying method; he preferred poison. "Much more effective; arsenic is good, strychnine is ideal. Let us put poison everywhere!" Then Father Fontanel provided his own personal tally in this "scientific hunt": in August 1918, he killed 3,157 English sparrows. His best day, he reported, was August 16, when he sent 435 birds to their Maker: "I buried them myself," he wrote, "so I'd be sure they were dead." By 1924, he felt he had perfected his plan: that year, he shot, strangled, trapped, or poisoned a total of 15,000 birds. "There can be no better proof," he wrote, "of the efficacy of my methods." If it is true that God sees the little sparrow fall, then He must have been keeping a close eye on Father Fontanel.

Many other efforts were undertaken to control house sparrows. Farmers began organizing "Sparrow, Rat and Crow Campaigns." The newly formed Audubon Society quietly supported these measures by regularly reporting in its bulletin the efforts of its correspondents to rid the country of this turbulent pest by means of traps, poisons, and firearms, usually to no lasting avail. Sparrow Clubs were formed, which held regular shooting matches with prizes such as free theatre passes for the most sparrow heads turned in. The clubs also tried for a while to have live house sparrows substituted for live passenger pigeons at trap-shoot meets. They encouraged the fashion for stuffed songbirds on ladies' hats, although birds more colourful than the house sparrow were usually the victims of the millinery trade (again, the eastern bluebird was the chief sufferer). One of the more interesting measures was an attempt to introduce house sparrows as a food item, on the "if you can't beat 'em, eat 'em" principle. In 1887, farmers at the St. Lawrence Market in Toronto paid young boys with guns $1 per hundred for house sparrows that, a short decade earlier, had cost bird-releasers $1 each to import. Sparrows were served in restaurants under the names "rice-birds" or "reed-birds," as substitutes for the declining bobolink, which had been commonly eaten under those names in the south. House sparrows, reported the *Express* of Albany, New York, in November 1887, "make excellent pot-pie and are regarded as excellent eating by those who have made the trial. The flavor is said to be somewhat like that of reed-birds and much superior to quail."

Although efforts to control house sparrows enjoyed local and short-lived success, the species continued to increase until the 1920s, when it began suddenly to decline. In the end, its decrease had less to do with Sparrow Clubs than with the gradual disappearance of horses – and therefore seed-rich horse droppings, upon which house sparrows had come to rely for their winter foraging – from city streets. This development was reported with no small satisfaction in the popular journals. "Among the permanent residents the English or house sparrows are decreasing in numbers," wrote I.S.E. Lafferty in the *Canadian Magazine* for November 1924. "This is more noticeable in towns and cities where horse-drawn vehicles have been largely supplanted by motor-cars. The English sparrows are husky, hearty chaps, but with their food supply diminishing the number of these 'rats of the air' must necessarily diminish also. Shrinkage in the sparrow population sounds like good news, and still more delightful is the report that the cardinals and redbirds are increasing numerically."

But despite the efforts of Father Fontanel and the proliferation of the automobile, the "husky, hearty" little house sparrow has refused to die. It has, in fact, adapted well, with some unintentional help from us. Precise statistics are lacking, as we say when we are making the wildest guesses, but I'd be willing to bet that there are as many bird feeders in the city now as there were horses in 1940, at least during the winter. My evidence for this is that there are about the same number of house sparrows in Toronto now as there were then, and food availability is a better regulator of wildlife populations than shotguns. By far the most frequent visitors to our feeder are house sparrows. They come in groups of six or eight or ten, always in pairs. They perch on the clothesline or the fence, jerk their heads around seven or eight times a second on the supposed expectation that some super-cat is going to come hurtling at them from behind our garden shed and swallow them all with one gulp. When that doesn't happen, one of them flutters up to the feeder and proceeds to use its beak to shovel great quantities of distelfink out of the feeder and onto the ground below, which happens to be, in summer, the bed in which we plant our salad greens. Then all the other house sparrows leave off their twitchy look-out for predators and fly down to the ground to begin pecking away at the spilled seed. For a long time I puzzled over this behaviour, why such a cautious, not to say paranoic, creature would rather pick its food from the ground than crunch it comfortably in the relative safety of a feeder hanging three metres higher than even a super-cat could leap. But house sparrows are, after all, weed-seed eaters, and weed seeds grow low and, when ripe and most fully charged with nutrients, fall to the ground. That's the evolutionary niche these "rats of the air" have chosen to occupy, and no amount of my moving their food up onto a tree-high platform is going to make them change their ways — why would they want to compete for some other bird's airspace? By tipping the bird seed onto the ground and pecking at it down there, they are turning it into the food they have already adapted to once: horse buns. Let them have it. I don't really know what distelfink is, but I hope it tastes good with arugula.

When they've eaten their fill or else can't stand the strain of exposure any longer, they fly abruptly away; first one cracks, and then the others follow suit, on the off chance that the first one was right. Some of them make a dash to the front of our house, where there is a medium-sized cedar tree they can hide in. On the morning after I'd watched the two sparrows cleaning

out their nest in the grocery-store wall, I happened to notice
that the snow at the base of this cedar was littered with the
tiny ends of cedar branches. Fluttering about in the shelter of
the tree – on the south side, of course, where the sunshine was
warmest and the new growth on the tree was strongest – were
a pair of house sparrows. Every few seconds they would take
the end of a cedar branch in their beaks, give it a sharp twist,
peck off the new growth, and let the rest of the frond fall to
the ground. They must have been coming there for days to
bask in the warm spring sunlight, feast off the pungent new
cedar buds, and then flutter off to finish their spring cleaning.
I'm pretty sure the cedar tree can withstand the onslaught, and
a new lining of crushed cedar buds in last year's nest will no
doubt help keep the lice and mites at bay.

If Colonel Rhodes was influenced by that fictional poem
by Jean Lesquillon, then the world's most successful introduction of a foreign species owes a lot to poetry. Damase Potvin,
who wrote an account of the early struggles of the house sparrow in *La Revue de l'Université Laval* in 1953, doesn't say, and perhaps we'll never know. But if the good colonel did know
about the poem, and did believe it to be a true story, then
chances are he never read the final stanza. Damase Potvin
quotes it at the end his account, and a rough translation of it
seems a fitting way to close mine:

> Two lessons from this tale I pluck
> And may they be our teachers
> *Primo*: All our surplus truck
> Should feed our smaller creatures.
> *Secundo*: God's eternal laws
> Have their own nomenclature;
> Until we understand their cause,
> Don't mess with Mother Nature.

SNAKES
IN THE GRASS

I am standing on the main road that runs along the spine of Leslie Street Spit, looking across a low field towards a pile of concrete rubble. It is a brisk Sunday morning in late April. The sun rose about an hour ago. There are sheets of water shining in the field, and knee-high couchgrass, and the bare branches of shrubs. Perfect snake habitat, I am saying to myself. I'll bet if I walk across this field towards that pile of broken concrete, I'll find some snakes. Yup, I'll just bet there are plenty of snakes in that pile of old, broken-up concrete. All I have to do is walk across this boggy field, through the tall grass, and over to that pile of broken concrete, and I'll see dozens of snakes out basking in the morning sun, slinking around in the grass catching frogs and swallowing them whole, in that slow, deliberate, inevitable way snakes have, crawling out of the crevices and niches in that pile of broken concrete, out of the darkness, their tongues flicking, their eyes unblinking, looking around, seeing everything, maybe slipping out of their skins. I take a deep breath and look down at the ground at my feet. I step into the tall grass.

At this point, I am not a big fan of snakes. I am not exactly afraid of them, it's just that ... well, all right, I am exactly

afraid of them. Seeing a snake sets off a flight response in me
that is so deeply ingrained it must be one of those involuntary
responses that define me as a *Homo sapiens*, like blinking when
something flies at my face, or holding my breath when I go
underwater. Fear of snakes is a species trait, like an elephant's
fear of mice. Snakes, like floods, appear as the bad guys in so
many of the world's mythologies that it seems disingenuous
to maintain, as E.B.S. Logier does in *The Snakes of Ontario*
(1958), that "fear of snakes is not natural in children." According to Logier, recoiling at the sound of something slithering
through grass is a response learned from herpaphobic parents, "directly by teaching, or indirectly by example." My
own daughters were wary of snakes without my having to
teach them to be; they just didn't automatically translate wari-
ness into hatred, as most adults do (and not just with regard
to snakes). Saying that fear of snakes is an acquired characteristic is really to suggest that we all lived perfectly comfortably
with snakes until we read Genesis and found out that creatures that crawled around on their bellies and ate dust were
Satan's get. It is far more likely that whoever wrote Genesis
wrote the snake into the Garden scene to explain a perfectly
natural phenomenon, which is what myths do and why they
work: we have an instinctual dislike of snakes. Myths do not
cause us to behave in certain ways; they simply try to explain
why we behave the way we do.

Myths can also be interpreted in different ways. Bernard
Arcand, a cultural anthropologist at Laval University, points
out that in Genesis, God lies to Adam and Eve, and the serpent
tells the truth. When God warns the couple not to touch the
forbidden fruit, he tells them clearly that "in the day that thou
eatest thereof thou shalt surely die" (Genesis 2: 17). The serpent, however, tells Eve not to worry about God's prediction:
"Ye shall not surely die" (3:4). Who turns out to be right? "As
everyone knows," writes Arcand in *The Jaguar and the Anteater*,
"first Eve and then Adam give in to temptation. But what is less

often noted is that everything that happens after that tends to confirm the predictions of the serpent." Despite such concrete biblical evidence, however, we still don't trust snakes.

Later stories capitalize on our natural disinclination. Dragons are nothing but big snakes. Sea serpents swallow ships. To me, the most memorable passage in *Huckleberry Finn* is the one in which Huck puts a garter snake in his teacher's pencil case, and then *puts the pencil case back in her desk drawer!*

The assumption seems to be that all snakes are poisonous – when Huck Finn's teacher puts her hand in her pencil case to get a pencil, she's going to be bitten by that snake, even though we know it's a garter snake and about as dangerous as a kitten. There's an Okanagan legend, called "Why Garter Snake Wears a Green Blanket," in which Garter Snake wins a fire-spitting contest with Thunder-Bird, spitting "a stream of sizzling fire that flashed right in the monster's face." Garter snakes do not spit anything, let alone fire, dragon legends notwithstanding. The Bible contains several references to fire-spitting serpents – Isaiah, for instance, speaks of Egypt as "the land of trouble and anguish, from whence come the young and old lion, the viper and fiery flying serpent" – and they have been taken as references either to the cobra, which spits into its victims' eyes before striking, or to some other venomous snake of the Egyptian desert, the bite of which stings like fire. Now, garter snakes are not venomous, as the Okanagans well knew, and yet the legend of Garter Snake and his fiery spit clearly plays on our assumption that snakes are bad news: don't mess with them. They are also treacherous, sneaky. In fact, the first known use of the verb "to sneak" was by Shakespeare, in *Henry IV, Part I*, when Hotspur describes the king as "Sicke in the Worlds regard, wretched, and low, / A poore unminded Out-law, sneaking home": Shakespeare almost certainly meant "snaking" home.

Our attitude towards snakes comes in part from their weird shape and method of locomotion. There are very few body plans on our planet: for all our talk of biodiversity, the Earth's bio isn't all that diverse, at least structurally. Head at one end, tail at the other, four legs in between pretty much describes all the vertebrates, from amphibians to fish to birds to orangutans to dinosaurs. Take an X-ray of an adult frog and hold it beside one of a three-month human fetus, and you'd have a hard time saying which is which. When human beings gaze on just about any other creature in the biological spectrum, they see something that more or less resembles themselves in body pattern. They see something familiar. They know where to look for, say, the tibia. This may be why human beings who design space creatures for science-fiction movies usually come up with something resembling human beings: most things in our known universe follow the same pattern, so why shouldn't most things outside it do the same? The really successful sci-fi films, like *Alien II*, are those that have creatures from outer space that look like giant crickets, not ones in which the aliens are single-celled amoeba that are threatening to alter the chemical structure of, say, water, which I would find much more frightening because it is much more possible.

Well, snakes don't resemble crickets. Snakes are the only land vertebrates in the known universe that don't have legs. Even whales have legs, at least vestigial ones: every now and then, a whale is born with two little legs sticking out the back in exactly the place you'd expect to find them. Not snakes. Some larger species of snakes – boas and pythons – have vestigial pelvises into which legs must at one time have fit, but our snakes don't even have those. At some point in their history, snakes realized that slithering required less energy than walking on whatever stumpy legs evolution had given them (alligator legs are incredibly inefficient for walking on land), or gave them an advantage when hunting, and so, being unfathomably intelligent creatures, they just evolved legs out of their body plan. A snake's skeleton today is basically a backbone and ribs with a skull attached to it. All the usual organs are in there – lungs, kidneys, liver, stomach, intestines, etc. – neatly arranged in a compact, narrow tube, as if packed by a very experienced traveller. But no legs.

Uncannily, leglessness doesn't slow them down a jot. In fact, snakes move with disconcerting quickness (up to six and a half kilometres an hour), although they don't appear to be moving at all. They do this in several mysterious ways. Snakes have plates, called ventrals, running across their bellies, with each row of ventrals controlled by a separate muscle attached to a rib. These muscles move in waves along the length of the snake's body, alternately bunching and expanding the ventral skin, so that the body is propelled forward. Little of this effort is visible from above, however, and so to such wide-eyed observers as ourselves the snake appears to be floating along the surface of the earth like a disembodied ghost, especially when the snake's body is in the "S" position, when its forward motion is enhanced by its being able to push sideways against the ground. The letter "S," by the way, is thought by some linguists to have derived from early depictions of snakes, maybe as signs painted at the entrance to caves: "Don't go in there!"

"Fear of snakes is a perfectly understandable primate response," says Bob Johnson, the Metro Toronto Zoo's reptile curator and author of *Familiar Amphibians and Reptiles of Ontario.* "Human beings evolved in semitropical zones, where most snakes *are* dangerous, and so the proper primate response to the presence of a snake is to recoil from it." Johnson adds to this the fact that, in southern climates, bright colours such as red and yellow are usually warning signs meaning Don't Touch; since many snakes in our region have red and yellow markings, our reactions to them become more understandable. Johnson gets about a call a day at his office at the zoo, mostly from people who say they have a snake in their back yard and want to get rid of it. "First I say that there are no venomous snakes in Toronto, and second that the best way to get rid of the snakes we have is to plug up their holes and remove their food sources."

Most snakes move into people's yards to catch prey and move out again when the prey does. It's true there are no venomous snakes in Toronto – the only vipers in Ontario are the massasauga rattlesnake (*Sistrurus catenatus*), which is found along the Detroit and St. Clair rivers, and up the shore of Lake Huron to Manitoulin Island, but not in Toronto, and the timber rattlesnake (*Crotalus horridus*), which used to be common as far north as Fitzwilliam Island, and may still be found in the Niagara gorge, although it hasn't been seen for a number of years. Toronto's snakes are all benign. The eastern hognose snake (*Heterodon platyrhinos*) will sometimes mimic a poisonous snake when it feels threatened. A thick-bodied, light-brown snake that grows to a metre in length, with dark blotches along its back and a pointed, upturned nose useful for burrowing through sand in search of toads, upon which it feeds almost exclusively, the hognose is rare in Ontario and even rarer in Toronto, but some have been reported in the sandy areas of High Park. When cornered, the hognose will coil up like a rattlesnake, raise its head off the ground, and flatten its neck to twice its normal width, so that it looks like a cobra – which is probably why it is also called a puff adder or a sand viper – then hiss in a most convincing snake-like manner. It may even strike (though keeping its mouth closed to avoid damaging its jaw) in an attempt to scare off its assailant. If further threatened, however, it flops down on the ground and proceeds to writhe around, mouth open, forked tongue lolling out, apparently in the final throes of acute ptomaine poisoning. It then gives up the ghost in a final paroxysm of agony, rolls over onto its back, and remains perfectly still and limp; if it had hands they would be clutching its breast. Even if you pick it up it will hang spinelessly in your hand, and if you flip it over on its stomach, it will roll back onto its back: "Apparently," writes Logier, "from this snake's point of view –

if it has any — a dead snake should lie only on its back." Walk away from it a short distance, however, and it will slowly raise its head and look around, and then slither off into the underbrush.

Hognose snakes have two characteristics needed for their highly specialized diet. One, shared with many reptiles, is a resistance to toad toxin. Many toads have glands in their heads (those large lumps behind the eyes) that secrete what is known as "bufo toxin" — *bufo* being Latin for toad — a white, milky fluid that, when ingested, causes mammals to gag and foam at the mouth. The hognose is immune to it. Its other adaptation is a pair of large, sharp teeth set well back in its mouth: when attacked, a toad puffs itself up to about twice its size, as if to fool the snake into thinking it is too big to be swallowed even by something that can unhinge its jaws. The hognose uses its fangs to puncture and deflate the toad and then proceeds to swallow it in the prescribed manner.

Actually, there is no prescribed manner. It is sometimes maintained that snakes always swallow their victims head first, but observation has shown otherwise. I have come upon a snake with a half-swallowed, live leopard frog sticking head-first out of its mouth, the frog's face looking resigned and expressionless as it took its last look at the world. Snakes do not actually unhinge their jaws: each side of the lower jaw is joined in front by an elastic integument, and both sides are attached to the skull by a linkage of movable bones. The jaw itself is equipped with a row of inward-pointing teeth, like barbs, which the snake fixes into the toad. It then wraps its mouth over the toad's body and proceeds to haul the toad down its throat by retracting first one side of its jaw and then the other, until the whole thing is worked down into its gullet, a process that can take from fifteen minutes to an hour. The opening to its windpipe (the glottis) is located at the base of its tongue rather than at its throat, which allows the snake to breathe while swallowing, like a teenager, and to hiss loudly while sticking its tongue out at you. The tongue, by the way, is actually a nose: scents adhere to it, and the tongue delivers them back into a cavity in the roof of the mouth called Jacobson's organ. The faster a snake flicks its tongue, the more interested it is in the smells it is detecting.

The eastern milk snake (*Lampropeltis triangulum*) is more common in Toronto than the hognose, but probably less frequently seen, being nocturnal. A long (up to a metre), tan-coloured snake with dark-brown blotches on its back and sides, the milk snake will sometimes attempt to defend itself by imitating a rattlesnake – it will coil up and vibrate its tail when alarmed, and is also called a "hardwood rattler" because of it. I once saw one on a quiet country road in broad daylight, and when I approached it cautiously it simply lay there like a limp lump until I moved away. I didn't threaten it, it didn't threaten me. Rodents form 70 percent of the milk snake's diet: it kills them by constriction, like a boa but on a somewhat smaller scale. It acquired the name milk snake from a popular belief among early homesteaders that it sucked milk out of cows' udders, but this was just another one of those stories we tell about snakes: it did hang around barns, apparently, but then so do mice, the milk snake's principal food. Toronto's populations of milk snakes still tend to be found where there used to be old farm buildings, but these days it's probably because old farm buildings have thick, rubble-stone foundations that make perfect hibernation sites for milk snakes.

Good habitat is only one advantage that the city offers to snakes. Another is a more benign climate than that found immediately outside the city. "Winter is the most critical time in the life history of cold-blooded creatures," says Johnson, "and winters in the city tend to be less severe." The northern brown snake (*Storeria dekayi*), also known as DeKay's snake, is one species that enjoys Toronto's microclimate, especially in the fall. Most snakes disappear at the first signs of cold weather in September – certainly the frog- and toad-eating snakes do, because that's when frogs and toads go into their hibernation pools. But brown snakes can be found sunning themselves on bicycle paths in Toronto's ravines well into October, when the temperature dips to 0°C. A small, stout snake, rarely reaching more than thirty-five centimetres in length, its colour varies from light brown to dark or greyish brown, with a light-coloured stripe down the middle of its back flanked by two darker stripes and two rows of dark, zig-zaggy spots. It, too, is nocturnal, hiding out during summer days under pieces of discarded cardboard, roofing shingles, or sheets of metal siding, and coming out at night to feed on slugs and earthworms. Its preferred urban habitat is in vacant lots, parks, gardens (for the slugs), lawns (for the earthworms), and the waste spaces between warehouses. The young, born live in August, look a lot like earthworms themselves and are often eaten by birds, and sometimes by milk snakes. Brown snakes never bite and, according to Logier, "make an attractive pet, becoming very tame and doing well in captivity." They are the consummate city snake.

The most common snake in Toronto, however, is the eastern garter snake (*Thamnophis sirtalis*). Its black or olive-brown body with its three bright yellow stripes is what most of us think of when we think snake. The stripes are its camouflage, often found on animals whose response to a threat is to flee. Try to focus on a rapidly slithering garter snake, and your eyes become confused and your mind disoriented; if you were a predator, you'd be unlikely to score a direct hit on any vulnerable part of it. Animals that freeze when alarmed tend to have spotted or "cryptic" body patterns, so that their own camouflage blends in with their surroundings.

The garter snake is the first snake to appear in the spring and, except for the brown snake, the last to go in autumn. It emerges in late March and generally hangs around until the first rains in October. Its preferred habitats are the grassy, open spaces near water, or along roadsides that have ditches; it eats earthworms, leopard frogs, young toads, small fish, nestling birds, and sometimes newborn mice. Contrary to popular belief, snakes do not dig holes of their own, but are often seen crawling out of rodent burrows, licking their non-existent chops. Garter snakes can grow to a maximum length of 124 cm, but their usual length is about 90 cm. Young garter snakes shed their skins about every two or three weeks as they grow. Actually, what they shed isn't really skin (their skin is under their scales, as it is in fish); they shed an outer coating made of keratin – the same stuff their scales, birds' feathers, and our hair is made of – which protects their scales and even their eyes from being damaged as they crawl about among sharp rocks and twigs. Shed skin, or slough, is often picked up by birds to be used as nesting material, either because it is very soft, or else because its scent keeps predators away from the nest.

When cornered, garter snakes will curl up, raise their heads above their coils, hiss, and vibrate their tails like rattlesnakes, and will even try the closed-mouthed strike to ward off attack. It has been suggested that they have learned these tricks from rattlesnakes, with whom they often share habitat, but it is more likely that their response to a perceived threat is as instinctive as ours is to them. Garter snakes have a secondary line of defence: when handled, they secrete a foul-smelling, sticky paste from their cloaca – the fecal vent at the spot where their stomach becomes their tail. No one has yet suggested that they learned this trick from skunks, with whom they also often share habitat.

In the fall, garter snakes migrate en masse to communal hibernation sites called "hibernacula." The most famous garter snake hibernaculum is in Manitoba, near the town of Narcisse, where an estimated 8,000 red-sided garter snakes (*Thamnophis sirtalis parietalis*) come together each fall, some of them from great distances. There are several sink-holes in salt deposits used by red-sided garters as hibernacula in Wood

Buffalo National Park in northern Alberta. The idea is to get below the frost line. Toronto has its own hibernation sites, subterranean mazes of piled stones such as are found in the foundation walls of old buildings, or in the rocky roadbeds under railway tracks, or, the best sites of all, those squares of white rocks wrapped in chicken wire, called "gabion baskets," we use to create fake stream-banks in our ravines and parks. The Metropolitan Toronto Conservation Authority has built two hibernacula in Colonel Sam Smith Park, a landfilled peninsula not unlike Leslie Street Spit, in Etobicoke, to attract garter snakes to winter in them. Every fall, garter snakes migrate to these sites along specific paths, travelling up to nine kilometres and sometimes taking several days to make the pilgrimage. Migration in any animal raises a few interesting questions: how do they know when to migrate, and how do they know where to go? Like birds, snakes follow the same paths to the same wintering grounds year after year. In the city, these paths are often along such urban corridors as railway tracks or ravines. A study conducted in Wood Buffalo National Park showed that garter snakes use shifts in the length of day to decide when to migrate towards the hibernaculum in the fall, and, upon emerging in the spring, use the sun again to determine the direction of their habitual feeding grounds.

The males emerge from underground gradually in late March or early April, coming out for a few hours each day to bask in the weak sunlight, building up energy for the mating frenzy that ensues when the females wake up a few weeks later. Lipids in the skin of the females emit pheromones that are detected by the flicking tongues of the males – these pheromones were isolated by researchers at the National Heart, Lung and Blood Institute in Washington, D.C., from garter snakes collected in Manitoba, and were the first to be found in vertebrates, although their existence had been suspected since the 1930s, when the role of pheromones in sexual stimulation in human beings was first discussed (about the

same time we started using under-arm deodorants). As the
female slithers sleepily out of her hibernacula, she is quickly
surrounded by up to a hundred primed males, each of which
entwines himself around her to form what is known as a mating ball. As the female warms up, she becomes receptive to
one of the males, who wraps his tail about hers and unsheaths
his hemipenis, which until now has been secreted behind the
ventral scales immediately below his cloaca.

The hemipenis is an oddly shaped organ: like the snake's
tongue, it is forked, with each prong of the fork shaped like a
stack of circular barbs. The male inserts one prong into the
female's cloaca, keeps it there by means of the barbs until he
has ejaculated, then withdraws it and inserts the second
prong, and ejaculates a second time. After this second ejacu-
lation, he introduces a mating plug – probably made up of the
male pheromone squalene – into the female's cloaca, as a
chemical message to other males that the female is no longer
available for mating. The other ninety-nine males slope obe-
diently off in search of more receptive females.

This may seem a dubious reproductive strategy – a huge
ball of frenzied snakes paying no attention whatsoever to the
possible interest of predators such as skunks, American
kestrels, and red-shouldered hawks, all of which have been
known to swoop down on unsuspecting snakes. But the success of garter snakes in cold northern climates attests to its
effectiveness. The male's sperm can live inside the female for
up to five years, which means that a female can continue to
bear young even if she remains unmated for four years out of
five. "With thirty live young born each year, and the ability to
store sperm from previous years," says Johnson, "garter snakes
are well adapted to establish themselves in new environments
or to survive in disturbed ones."

Garter snakes were the first reptiles to appear on Leslie Street
Spit. They come in many colour variations: some have bright

yellow stripes, others have more muted stripes, still others are downright dull and sometimes even spotted. All these variations can be found in a single litter: evolution seems to favour wide variation in some animals, perhaps to ensure the survival of a species that has spread itself out over a wide variety of habitats. Look at *Homo sapiens*. The garter snakes that turned up on the Spit were completely black, with small white patches under their chins. This mellanistic, or black, strain was introduced to the Toronto Islands in the late 1980s, from a colony near Point Pelee, and it appears that some have made it across the harbour to over-winter in the man-made rubble that forms Leslie Street Spit. I was hoping I might see some of them. Sort of.

"You almost never find snakes when you're looking for them," Bob Johnson had warned me. "The best thing to do is to go out looking for something else – birds, for example – and then you won't be disappointed if you don't find snakes."

So as I walk across the grassy field towards the pile of broken concrete, binoculars bouncing from my neck, I look around for birds. There are plenty to be seen, even though spring migration is still a week or so off. The bushes harbour a few palm warblers and a savannah sparrow, and there are white-throated sparrows hopping around on the ground at the base of a thicket of alders. A flicker skims the tips of the grass and briefly lands on a low maple branch before disappearing down towards the shoreline. I am walking steadily towards the pile of broken concrete, my shadow thrown along the brown grass ahead of me, when a pair of golden snipe start up about a couple of metres from my path and fly off in all directions. Their sudden panic scares the bejesus out of me – I thought it was a snake.

THE POST-COLONIAL
TERMITE

I n many termite-infested cities, the arrival of spring is her-
alded by millions of winged termites swarming out of their
underground termitaries, setting out on mass maiden voy-
ages to find mates and start termite colonies of their own. Unfortunately, Toronto isn't one of those lucky cities. It's not that
Toronto doesn't have termites – quite the contrary. It's that
Toronto's termites don't behave the way most other termites
do. And that, according to Tim Myles, is bad news.

Tim Myles is listed on the University of Toronto's registry
as the director of the Urban Entomology Program, a branch of
the Forestry Faculty. "When you're the only one in the program," he says, "they have to call you director." A young, mild-
spoken man with brown sand-stone hair, neatly trimmed
beard, and a finely tuned sense of irony, Tim is to Toronto's
termites what Buffalo Bill Cody was to bison. His office is a
narrow room on the second level of the Forestry building, part
of a complex of chambers and corridors radiating out from a
central stairwell that gives up to a skylight and down to a
ground-floor lab. The Forestry building itself is part of an even
more complex arrangement of edifices and colleges that make
up the university, and Tim shares his workspace with a computer that connects him through Internet with a wide web of

scientists and researchers the strands of which encircle the globe. On his desk are a stack of urgent departmental memos, a pile of back issues of *Termite Tips* — a desktop newsletter he produces for people with either a professional or a lay interest in social insects — and a coffee cup filled with an array of pencils, pens, and felt-tipped markers, two red and one blue. He came to Toronto from the University of Arizona in Tucson, and before that received his training as an urban entomologist at the University of Hawaii. Since it is safe to say that there are more termites per cubic metre of wood in Hawaii than anywhere else on Earth, with the possible exception of downtown Toronto, I don't have to ask him where he developed his interest in termites, or why he decided to come to here.

"Termites don't belong this far north," he says. "They are a tropical order, and if it weren't for us, they wouldn't be here at all." A termite's chief requirements, after wood or cellulose, are warmth and moisture, and since in climates like ours it is difficult to get from one warm, moist place to another without going outside, most termites restrict their activities to locales within chomping distance of the tropics. Only one species is found north of Indianapolis: the eastern subterranean termite, or *Reticulitermes flavipes*, the kind we have in Toronto. Subterranean termites build vast, underground galleries — just how vast is one of the things Tim is in the process of determining, but so far he knows it is "almost unimaginably vast." The outlets from these territaries lead directly into food sources, such as pieces of damp, rotting wood lying or standing on the surface. "Tree stumps are the best," Tim says, slipping not uncharacteristically into the termite's point of view. "They get into the root systems and work their way up into the stump itself. That way they never have to go out into the open air. And every yard in Toronto has a couple of tree stumps. Woodpiles are good, too. And fence posts. Playhouses. Picnic tables. Porches."

Termites don't exactly eat wood; they eat the cellulose that exists in wood. Depending on cellulose for one's total nutritional intake is a poor evolutionary choice, because the gut cannot digest it (as Darwin noted, "The gastric juices of animals do not attack cellulose"), and because even if it could, cellulose is a completely odourless, tasteless, and innutritious substance. It is what is left after wood has been boiled in water and alcohol for a few hours. In order to process it, termites had to find a way to compost it inside their guts. They did this (some 200 million years ago) by first ingesting protozoa, one-celled organisms that formed colonies inside each individual termite in what has been described as a perfect example of sym-biosis. The species of protozoa inside termites (each species of termite has its own species of protozoa, which they pass on from generation to generation by regurgitatory and fecal feed-ing) are now unique to termites: many of them are not found anywhere else in nature. Termites could not live without their minute colonies of protozoa, because they could not digest cellulose, and the protozoa could not live outside the termite, because they would have no means of acquiring cellulose.

Processing cellulose produces a lot of gas, mostly methane and carbon dioxide. A few years ago, it was thought that, worldwide, hundreds of billions of termites composting whole forests inside their guts produced 30 percent of all the methane that entered the Earth's atmosphere (with most of the rest coming from rice paddies and burping cows). Termites were then loathed not only as home-wreckers but also as polit-ically incorrect producers of a major greenhouse gas. It is now believed that almost all the gas produced by termites is carbon dioxide – workers breaking into termitaries in the tropics have been known to pass out from the sudden rush of carbon diox-ide – and the rest consists of only about 2 percent of the Earth's total methane budget. But that's still a lot of methane, and since carbon dioxide is also a greenhouse gas, termites are not off the hook as contributors to global warming.

To give termites their due, they do try to keep most of that gas to themselves, sealed in their termitaries. Whenever they absolutely have to cross metal or concrete to get from their galleries to a food source, they don't just troop across through the open air, because if they did they would dry up and die. Instead, they build "shelter tubes," long, hard, grey tubes made of earth glued together with their own dung. These tunnels can be seen snaking across sidewalks, up foundation walls, even up tree trunks. Termites travel through these connecting tunnels, back and forth from their house to yours, not unlike human commuters shuttling along the TTC from apartment building to office complex, without ever having to expose themselves to sunshine or air. Shelter tubes are one of the adaptations that have allowed *Reticulitermes* to move north: of the Earth's 1,802 species of termites, *Reticulitermes* is the only genus that lives naturally north of the Equator, and they do so in a 1,660-km-wide swath all the way around the world.

Toronto, you may say, is more than 1,600 km north of the Equator, and yet *Reticulitermes* is quite well established in, or rather under, our city. That is because of a few extra adaptations developed by Toronto's population of subterranean termites, adaptations that allow *Reticulitermes* not only to survive this far north, but to flourish in numbers that boggle even Tim's sympathetically inclined mind.

"In the south," he says, "in places like Florida, it used to be thought that colonies of *Reticulitermes flavipes* lived in subterranean colonies of about a quarter-million. That figure was based on some totally pulled-out-of-the-air assumptions about how far a termite could forage. Some guy sitting at his desk saying, 'I don't think they could possibly forage more than twenty metres, so if I find termites over there they must belong to a different colony.' A totally ridiculous assumption; it's no trouble at all for a termite to travel twenty metres underground. But that's where it stood for a long while."

In fact, that's where it stood until about ten years ago, when researchers from the Urban Entomology Program,

working with then director Ken Grace, started doing mark-
release-recapture studies on Toronto's termite colonies. This
involved capturing hundreds of termites in specially designed
termite traps distributed throughout the Metro area: the traps
were sections of PVC pipe stuck in the ground and filled with
rolls of damp, corrugated cardboard: fast-food outlets for termites. The researchers fed each of the 10,000 termites they
captured on fat-soluble red dye, released them through the
same traps they were caught in, and then counted how many
red-marked termites they recaptured in the traps a week later.
From this, they could calculate how many termites there were
in each colony.

   Their calculations shattered all former notions of colony
size for *R. flavipes*. In fact, they shattered all former notions
about what a colony was. "We came up with estimates an
order of magnitude greater than the southern counts," says
Tim. "When they said colonies must exist in the hundreds of
thousands, we found they existed in the low millions. When
they said colonies must be about twenty metres apart, we were
finding them stretching across fifteen hundred square metres."

   Tim doesn't even talk about colonies any more. Using
another apt parallel to North America's human population, he
now talks about termites inhabiting, not individual colonies,
not even city blocks, but entire neighbourhoods, whole bor-
oughs. And in astronomical numbers. "We figure each city
block in the parts of Toronto that are infested represents
approximately 10 million termites," he says. "And we know
that there are 545 infested city blocks, or about one house in
fifteen." He leaves me to do the math as we descend the stairs
to his lab. I'm up around five billion termites when he adds,
"And that's just the houses we know about. That doesn't count
the areas where, for one reason or another, nobody wants us
to know they have termites."

   In the lab – a large, well-lit room filled with several ranges
of counters and the sound of running water – there are stacks

of plastic tubs, each containing a few layers of wet corrugated cardboard. Tim begins lifting the layers to show me that under the cardboard the tubs are about a quarter full of what looks to me like cooked rice – not cooked white rice like Uncle Ben's, but maybe cooked brown rice, or basmati rice cooked in a light soy sauce. When I look closer the rice seems to be moving, like cooked rice with legs and heads, thousands of them per tub, all scrambling – no, seething – to get out of the sudden light.

"These are workers, mostly," Tim says, poking at them with his finger, "but you can see a few soldiers – they're the ones with the orange heads."

Termites are the most social of the social insects. Like ants, bees, and wasps, termites exist in huge, highly organized colonies, so intricate in their structure that it hardly makes sense to speak of individual termites. In fact, in English the insect was originally called "termes" (the Latin word for wood-worm) when there was one of them, and "termites" (the Latin plural of *termes*) when there was more than one. Since there was almost never only one of them, the plural became the singular, and now a single termes is called a termite. The eminent entomologist Edward O. Wilson, who based his concept of the Earth's need for biodiversity on his life-long study of ants, speaks of a colony of ants not as a collection of individual insects but as a single "superorganism," the entire colony behaving as one entity. "One ant," he has said, "is no ant." He regards each ant colony as "an organism weighing about ten kilograms dominating an area the size of a house." Viewed in this fashion, a colony of termites might be seen as a 50,000-kilogram organism occupying an area the size of Scarborough. Tim goes even further than that. For the purposes of his research, he says, it no longer makes sense to study termites in a block of houses in Kensington Market, for instance, and call that Colony A, then look at another block of houses in Cabbage-town and call that Colony B, and so on throughout Metro

Toronto. "It might make more sense," he says, drawing his fingers across the entire map of Toronto, from the Scarborough Bluffs to the Humber Valley, "to call this Colony A and leave it at that."

Termites are often confused with ants – in the south, they are even called "white ants" – but there are several ways in which termites differ from ants, apart from the size of their colonies. Like ants, termites have two pairs of wings, but in ants the fore pair are much bigger than the hind pair. In termites, both pairs are the same size – hence their family name Isoptera, from the Greek *iso*, same, and *optera*, wings. Termites are soft-bodied, ants are hard-shelled; termites are light in colour and avoid direct sunlight, ants are either black or red and are often seen in broad daylight. An ant has a body like a snowman, distinctly divided into thorax and abdomen by a narrow waist, called a petiole; termites' bodies are not so obviously divided – they look more like cockroaches, and in fact are more closely related to cockroaches than they are to ants. And although termites and ants both swarm in the spring – that is, great hordes of them spread newly formed wings and fly into the air in April in a feverish search for mates – ants actually mate in the air before losing their wings and returning to their mundane life on the ground, whereas termites land first, shrug off their wings, and only then seek out a similarly unwinged mate. The male termite, having identified his future queen, follows her doggedly around until she enters some dark crevice in the ground, say where your porch steps meet your lawn, whereupon the two of them seal up the entrance to their new nest and do not emerge from it again. That, at least, is how textbook termites do it. Toronto's termites, however, have developed a few adaptational twists to the textbook theme.

Termite young hatch from eggs as nymphs, but do not (as ant nymphs do) go through the usual larva-pupa-adult phases; the newborns look like little adults and grow as snakes do, by

shedding their skin, or moulting. As they mature, they myste-
riously develop into one of the four termite "castes," or social
orders. Although ants also develop into castes, each ant caste
consists entirely of males – ant nymphs do not develop into
females unless one is needed to replace the queen. In termites,
all castes consist about equally of males and females. The
most common is the worker caste – the sterile, eyeless
drudges that groped in Tim's plastic tubs, whose job it is to
supply food to the colony and to build and repair the intricate
subterranean maze of tunnels and royal chambers that is the
colony's domicile. This is the caste that eats houses. It is a
caste of millions. Innumerable termite workers file in and out
of the subterranean galleries to seek and devour wood wherever it may be found, then return to the galleries to regurgitate it as food for the other termite castes.

The soldier caste, as the name suggests, guards the colony
from invasion by such predatory enemies as ants. Like workers, soldiers are blind and wingless, but they have elongated
orange heads and proportionately monstrous mandibles,
which are capable of seizing and crushing an invading ant if
one should blunder into a termite gallery by design or mistake. Soldiers can also exude, from an opening in their skulls,
a transparent fluid that is anathema to ants. (Ants, to maintain
a balance of arms, spray their foes with jets of formic acid,
which immobilizes them until they can be despatched by
more conventional weapons.)

The third and fourth termite castes are the reproductives,
and here is where Toronto's termites differ from their southern conspecifics. The fertile males and females with wings,
the "alates" (from the Latin *ala*, meaning wings), are the
springtime swarmers who mate and, in southern colonies,
move away to begin colonies of their own. Toronto's eastern
subterranean termites seem to prefer a less *al fresco* form of
reproduction. Most of our nymphs do not progress to the
alate stage at all, but rather "bud off" one moult earlier to

become "nymphoids" instead. Nymphoids (or neotenics) are, in Tim's words, "sexually mature adults with child-like bodies." These neotenics are the underground termites' secret weapon against the Canadian climate.

"All social insects produce an annual crop of winged, mature alates," says Tim. "Our termites do that too, but it isn't their primary method of colony foundation. Instead, what we're seeing in Toronto is that, at the second-to-last nymphal stage, the majority of individuals within the colony are shunted into nymphoids. They are child-like, except they are adults; they retain all their nymphal features, but their gonads are fully mature and they can reproduce. The second-to-last moult is the perfect time to do this, because on their last moult they develop the rest of the adult features – their compound eyes get bigger, their wing muscles get really beefed up, and their wings develop. All these are energetically expensive changes, and in a cold climate you don't want to expend a lot of energy unnecessarily. At the nymphoid moult, you're big enough, you've got lots of body fat, you've got all the calories you need, you can breed – you don't need all that expensive flight equipment. Why not breed right there in the colony and save all that energy?"

Which is what Toronto's eastern subterranean termites do. "I think we actually have a different species here," says Tim, "definitely a different subspecies, one that is following this strategy much more than the same species does in Florida. Here, the majority of individuals become nymphoids, and they don't exit the colony." In other words, in tropical termite species a colony has one queen that does all the reproducing for the colony, and the year's alates fly out to start new colonies elsewhere. In Toronto, our eastern subterranean termites have developed their own system, by which all the nymphoids remain within the colony, simply moving out to its periphery (which biologists call the colony's "suburbs") so that their offspring contribute to the size of the original colony. In addition, when the colony is under attack, all its members – the queens, the secondary reproductives-in-waiting, the nymphoids, even the workers – can turn themselves into reproductives and become automatic egg-laying machines in high gear, and you end up with a lot of termites. A single queen can lay several thousand eggs per day for life (and she lives about ten years): imagine what a whole population of desperate, egg-laying termites could produce if they felt threatened – say, by human beings pouring pesticides into their termitary.

It is hardly surprising, then, that traditional extermination methods don't work very well in Toronto. The usual way to get rid of termites in the tropics is to dig down into the colony, find and remove the queen, and then watch the rest of the termite colony trudge conveniently off somewhere else to start a new colony. This is what happens in Nadine Gordimer's short story, "The Termitary," which is set in South Africa: the narrator's mother hires three exterminators – a white Afrikaaner and two black helpers – to rip up her floorboards and find the royal chamber. "Somewhere under our house," Gordimer writes of the queen, "she was in an endless parturition that would go on until she was found and killed." When the exterminators find the queen, they present her to the narrator's mother in a cardboard shoe box: "We all gazed at an obese, helpless white creature, five inches long, with the tiny, shiny-visored head of an ant at one end. The body was a sort of dropsical sac attached to this head; it had no legs that could be seen, neither could it propel itself by peristaltic action, like a slug or worm." The men take the queen away and the rest of the termites disconsolately abandon the termitary. The death of the white queen and the ensuing mass exodus of her white subjects had powerful symbolic significance in South Africa during apartheid, but nothing of the sort would have happened in Toronto. In Toronto, the men would have dug and dug until they found and removed a white queen, and then, in response to the digging, some 10 million other termites would have become queens.

Termites were unknown in Toronto until 1938, when a colony was found in a warehouse on Cherry Street just south of the Keating Channel. The warehouse is gone now, and in its place there is a Public Works yard filled, appropriately enough, with concrete telephone poles. By the time city officials became aware that Toronto had termites, the warehouse had been infested for nearly a decade, which may explain why it is no longer there. My father-in-law, who grew up in the Danforth-Woodbine area of Toronto, remembers watching his own father jab a screwdriver into what felt like a soft floorboard in the dining room and pulling up nothing but a thin veneer of varnish. That was in the early 1940s. By then, there were serious outbreaks throughout the southeast section of the city, along Toronto Bay, east into Scarborough, with termites spreading like an underground bushfire into nearly every corner of Metro. No part of the city is without termite problems: from the waterfront to North York, and from Scarborough to Mississauga, nearly 18 percent of the city's residential and commercial buildings are being digested by termites, almost all of them eastern subterranean termites. Toronto's Housing Department has, since 1965, approved 7,550 houses for treatment in its termite control program – a small fraction of the number of infested houses, but a lot of houses nonetheless. The office gives "termite remediation" grants (the maximum has just been raised from $125 to $500) to homeowners who want to discourage termites by reducing wood-soil contact around their houses, but it's a rearguard action: the termites are laughing at us.

When Toronto writer Marni Jackson discovered termites in a huge Manitoba maple in her yard three years ago, she called the Housing Department to find out how long she had before her new back deck collapsed. The officer who went to look at her tree was Bob Lott. "Oh yeah," Lott told Jackson when he looked at her Manitoba maple, "they're having themselves a ball in there."

Tim Myles was with Lott because he was interested in the fact that Toronto's *Reticulitermes* population had also developed a unique propensity for climbing trees. Termites do not like living wood because the sap fills up their foraging holes, and so living trees are usually safe from their attentions. But Toronto's termites build shelter tunnels up the trunks of living trees to get at dead branches higher up. "Usually," says Tim, "they go up after a large pruning scar that's about six to eight feet above ground. I don't think it's random foraging, either," he adds. "I think they know there's dead wood up there. We know that fungus will grow on dead, wet wood, and in the spring the rains will wash a trail of that fungus down the tree trunk. And we know that wood fungus contains exactly the same chemical as is found in the termite's trail pheromone: essentially, the trail of fungus on the tree is saying, Hey, this way to the food! So I think they chemically detect dead wood at the base of the tree and follow it up to the pruning scar."

Marni asked Bob Lott if having termites in her yard meant she had termites in her house.

"Not yet," Lott told her. "First they'll kill your tree, then they'll eat their way towards the house. They can move an inch a day. You have to eliminate all wood-soil contact, then pump poison into the ground. What you want to do," he said, "is create a chemical moat around your house."

Pouring pesticides around your house is little more than deliberate soil pollution. Although it has been the standard procedure for treating termite infestations in Toronto for years, it is not much more effective than digging up the queen. In fact, chemical soil injection is in large part responsible for the huge number of termites now chomping their way through our walls and foundations. The chemicals kill termites on contact, one at a time, and termites are clever enough to stay away while the chemicals are in the soil. The compounds – polychlorinated byphenyls like dieldrin were used extensively until they were banned in the 1980s – do not hang around for long, but leach out into the ground water and end up in Lake Ontario. Pest controllers now use a milder chemical sold under the trade name Dursban TC, but the method and results are the same. The chemical moat procedure has been used so widely because, until recently, not much has been known about the life cycle of the eastern subterranean termite – Tim's researchers have only recently learned about the nymphoid adaptation, for example – and because, also until recently, it was the best control method anyone had come up with.

But chemical moats merely act as a deterrent, preventing the termites caught in the house when the ground outside is treated from returning to the colony (they will eventually dehydrate in the house and die, unless they find a moisture source inside the house walls – have you checked the floor around your shower lately?), and keeping the main colony out of the house until the chemical dissipates into the soil and leaches into the ground water. More often than not, the chemical moat approach doesn't even do that. "Any house has thousands of entry points for termites," says Tim. "You're going to miss at least one crack in your foundation wall, and 10 million tiny nervous systems are going to find it. Until then, they're still in the soil, they're still in the fence, they're still in the woodpile, they're still in the stumps. When you treat a house with a chemical soil pollutant," he says, "you're doing nothing to the population."

Tim believes that he and his researchers have come up with a better termite trap. The idea for it came while they were doing a series of behavioural studies, using those red felt-tipped markers Tim keeps on his desk to brand captured termites. The red dye had to have the right combination of solvents and resins to give a nice, nontoxic mark that stuck to the insects' waxy cuticle for at least two weeks. Termites, like most social animals, spend a lot of time licking dirt off one another, and Tim didn't want his experiment subjects to lick the dye off before the experiment was completed. It suddenly struck him that, in order to turn the marker into a perfect termite control mechanism, all he had to do was mix a termiticide – he chose the relatively safe chemical sulfluramid – with the ink before dabbing it on as many termites as he could capture.

"Termites have a tongue," says Tim, "called a hypopharynx, that sits right between their mandibles. It's much bigger in proportion to their head than our tongue is, and it's covered with lots of little setae, or hairs. When they groom other termites, these hairs get into the cracks and creases on the termite's cuticle, and clean out all the dirt and fungal spores, keeping the colony clean. And the grooming termite doesn't just spit all that stuff out – it swallows it." This behaviour, known as biological altruism (it's the same impulse that drives bees to sting intruders, even though it means death to the individual bee that does the stinging), is particularly interesting to entomologists because it says a lot about the dedication of the individual to the health of the colony. "If you put sulfluramid on a piece of wet cardboard," says Tim, "the termites won't touch it. But put the same poison in even higher concentrations on the back of another termite, and they'll lick it off and swallow it. They seem to say, 'It's awful, it's noxious, we know it's bad for us, but we've got to run it through our guts anyway to expose it to our gastric enzymes, and try to detoxify it.' That's part of cleaning." It's also part of keeping the colony alive, which ensures the transmission of each individual termite's genes into countless and ever-expanding generations.

What it means to termite control is that, instead of killing only the one termite that comes into contact with the chemical moat, Tim's trap-treat-release method kills hundreds, maybe thousands of termites for each termite treated. In tests run last year in the St. Clair Avenue – Jane Street area – one of the most heavily infested neighbourhoods in Toronto – Tim and his researchers were trapping up to 1.2 million termites per week from a single city block, dabbing them with the sulfluramid solution, letting them go, and coming back two weeks later to see how many termites were left: in every case, the termite population was almost completely wiped out.

So far, sulfluramid is not one of the chemical termiticides registered by Agriculture Canada, and so no one can use it without a special experimental permit. But applications for licensing are in, and Tim is confident that if all pest control officers switched to sulfluramid from the highly toxic and ineffectual insecticides they're using now, Toronto could be a termite-free city in five years.

Well, free of eastern subterranean termites, maybe. But Toronto has been visited recently by several new species of termites that pose new threats not only to the city's houses, but also to the furniture inside those houses. The new species belong to the general family known as dry-wood or powder-post termites and, as their name suggests, they don't need to invade the damp sapwood found in the foundations and unventilated wall cavities of most of our homes: they can live virtually forever in a series of self-contained galleries in small pieces of dry wood. Like sofa legs. Or pianos. Or the plywood backs of bookshelves. And they don't have to return to their underground galleries for moisture, as subterranean termites do. They manufacture their own water.

"That's why they're so dangerous," says Tim. "They produce all the water they need metabolically. When you oxidize cellulose, the by-product is carbon dioxide and water." Subterranean termites produce water this way, too, but for some reason they expel it. Dry-wood termites don't. "If you look at the rectum of a dry-wood termite," says Tim, with only a trace of a grin, "you'll notice that the strongest muscles in their bodies are their rectal muscles. They use these muscles to squeeze every last bit of moisture out of their fecal pellets, and they don't defecate until they have a completely dry pellet." In fact, their pellets are so dry that they are often mistaken for sawdust or wood powder – which is why they're called powder-post termites. "I hate that term, powder-post," says Tim. "Those little piles of tan-coloured sand you find under your Steinway piano are not wood powder at all, they're fecal pellets. Termite frass. The termites make a little kick-hole in the wood and eject these frass pellets out, then plug up the hole again with a doughy kind of paste." Trained observers look for these plugged-up kick-holes as signs of termite infestation: the rest of us just have to wait until the Steinway disintegrates into a pile of matchwood on the living-room floor.

The first dry-wood termite infestation was reported to the Toronto Housing Department in September 1989, by the owner of a house on Swanwick Avenue, in East York. A colony of western dry-wood termites (*Incisitermes minor*) had moved into the basement, probably from a piece of furniture ten years before. By the time the fumigators were called in – in June 1990 – the infestation was too severe to be treated locally, by drilling holes into the joists and injecting Dursban into them. Instead, the entire house was wrapped in plastic and pumped full of methyl bromide – a highly toxic gas that, like sulfluramid, is not yet available on the open market. Pest control workers say they saw the plastic tent fill up with termite alates just as they were turning on the gas, and Tim, watching from the sidelines, says the procedure was "a flawless eradication." It cost about $17,000, most of which was paid for by the province.

The following Monday, however, a Mississauga home-owner, alerted by media coverage of the East York fumigation, brought part of his kitchen cabinets in to the office of Geoff Cutten of the Hazardous Contaminants Branch of the Ministry of the Environment. From fecal pellets and a few immature termites in one of the boards, Cutten realized that he was looking at a second infestation of dry-wood termites, and he called Tim Myles. Here is Tim's account of what happened then, as recounted in the November 1990 issue of *Termite Tips*:

"Geoff and I visited the house, where we discovered that, in addition to the infestation in the removed cabinet, the termites had started a second colony in the floor beneath the first-floor bathroom. Large quantities of fecal pellets had accumulated on ceiling tiles in the finished basement beneath the bathroom. We removed many of the ceiling tiles and, by noting where the fecal pellets had fallen, we mapped out the extent of the infestation in the floor above. We even found two additional incipient colonies that had established in nail holes in the joists some distance away from the main colony. We meticulously swept up all the wings and pellets we could find. Later, I examined the pellets and found two fragments of soldier head capsules, which were enough to identify the species as *Cryptotermes brevis*."

Cryptotermes brevis, also known as the West Indies powder-post termite, is native to the Caribbean and Central America, but is also common in Florida, Louisiana, and Hawaii – the only other colony known to have settled in Canada came from Mexico in 1967, in a wooden tray: the termites escaped from the tray while it was stored in a cabinet in the department of zoology at the University of British Columbia, of all places, and are still there, reportedly alive and well.

So far, says Tim, dry-wood termites are not a menace in Toronto – underground termites still make up more than 95 percent of Toronto's termite population. Dry-wood species, he says, are "almost an academic curiosity." Three species have turned up – in November 1994, for example, a shipment of wooden crates arrived at Spar Aerospace in Mississauga, from Kuala Lumpur via Jamaica, and the plywood backing was found to contain a sizeable colony of *C. domesticus* as well as a smaller colony of *C. dudleyi*. The crates had obviously been made from wood that had been infested for several years – but a single sub-zero night outside the Spar Aerospace offices was enough to kill off both colonies.

But, Tim adds, this is not to say that dry-wood termites won't become a problem in the future. "There must be hundreds of houses out there with termites we don't know about," he says, almost happily. "Ten percent of all the furniture in the West Indies and Florida is infested with termites, so it stands to reason that at least that percentage of furniture imported from those places is infested as well. And Toronto is a very cosmopolitan city. These recent cases demonstrate that dry-wood termites are able to move from the original piece of wood in which they are introduced and establish successful infestations in houses. So it's only a matter of time, really. . . . "

When I called Marni Jackson to ask what she'd done about her termite problem, she told me she really hasn't done much except install a fifteen-centimetre metal strip, which she calls her "termite shield," between her back deck and the house to act as a barrier. So far, the shield has kept termites out of her house. "I figured I'd let them have the deck," she said, philosophically. But her neighbours' attached house is infested, and they have had their yard treated twice with Dursban, and now Marni feels morally obliged to have her yard chemically treated as well. "The other day," she said, "I found a shelter tunnel leading across the brickwork between our two houses, so I guess the writing is on the wall."

Her remark makes me think of Tim and his marker pens. "Or maybe *in* the wall," I say. She doesn't laugh.

It's almost as though Marni has decided she can live with termites, and in the long run that may be what the rest of us will have to decide as well. Any organism as determined to adapt and survive as *R. flavipes* will no doubt continue to find new ways to evade our attempts to stop them. "Eradication," says Tim, "means the total elimination of a pest from a given area. With introduced pests, this is rarely achieved. Instead, what we are seeing worldwide is the continual spread and

establishment of pest species – Norway rats, European starlings, Turkestan cockroaches, Africanized killer bees, the
zebra mussel, the list goes on and on." Tim finds this "homog-
enization of the environment" unsettling; in fact, he calls it
"an ecological disaster of planetary proportions." Introduced
species disrupt the natural ecological balance of a region,
often in the absence of natural predators. The result is a wide-
spread take-over of gigantic proportions. Japanese knotweed,
for example – a garden escapee introduced in California
thirty years ago as an ornamental – is now so pervasive it has
been called "the plant that ate the West." Tim thinks that a
species born and raised in Borneo should stay in Borneo; anywhere else, it's a pest.

I'm not so sure. It may be true that introduced species disrupt the native ecological balance, but disruption doesn't necessarily equal disaster, and the spread of a species into a new
ecological niche is the engine that drives evolution. It is how
one species evolves into a new species. Some day the Japanese knotweed that has taken over the West will differ from
the Japanese knotweed in Japan, and will be called California
knotweed, and the world will have two species where before
there was one. In charting the transformation of the eastern
subterranean termite into an organism that produces
nymphoids instead of alates, climbs trees in search of food,
and moves out to the suburbs instead of founding new
colonies, we may be witnessing the evolution of another new
species – a fascinating process that, in animals other than
insects, can take millions of years. We could regard the
opportunity as a privilege.

Besides, isn't it a touch hypocritical for *Homo sapiens* to con-
demn another species for moving into new territories, setting
up self-contained colonies, disrupting the local ecological
balance, and adapting to new surroundings? Why call it an
environmental disaster when *they* settle into new territory,
and an advancement of civilization when *we* do?

Summer

In intervals of dreams I hear
  The cricket from the droughty ground;
The grasshoppers spin into mine ear
  A small immeasurable sound.

— *Archibald Lampman, "Heat,"* 1888

GO EAST,
YOUNG BIRD

Birders get kicks out of Hollywood movies that are lost on most of us. Fred Bodsworth, the author of the novels *The Sparrow's Fall* and *The Last of the Curlews* and, incidentally, one of Canada's leading ornithologists, is telling me about a scene in *The Savage Innocents*, a 1959 docudrama about the negative effect whites have had on Inuit culture. The film has Peter O'Toole doing a voice-over and Anthony Quinn playing a young Innuk hunter. Much of it was shot near Churchill, Manitoba – the whalebone and sealskin kayak used by Quinn was still proudly on display in the Churchill Museum when I visited there a few years ago – but some of the exteriors were filmed in studios in California. In the scene Fred is describing, Quinn is out hunting on the barren tundra, in the spring, surrounded by permafrost and ruminating muskoxen. "I kept hearing these birds in the background," Fred says. "At first I thought, well, there are birds in Churchill. What kind are these? And so I listened – they were house finches! There are no house finches in Churchill, but there are plenty of them in California! I wondered why the hell they would go all the way to Churchill if they were shooting the outdoor scenes in Hollywood?"

Fred relates the story as we walk along a ravine near his house in the Beaches, looking for birds. He is taking part in the Toronto Ornithological Club's annual migratory bird count, and every morning in early May he spends an hour or two in the ravine, listening for and listing the number and species of birds he comes across. He has been doing this every year for nearly half a century, and there are few people in Toronto with a better ear for birds, or a better sense of the city's avian history, than Fred. In the 1960s, he made part of his living conducting birding expeditions around the world, taking paying groups to such high-profile birding spots as Cuba, Colombia, South Africa, and southeast Asia. His life-list is impressive to a novice like myself: "Oh, no," he protests, "only about 1,500 species. I was never much of a lister." I'd spent part of the morning thumbing through his bird books, coming across birds I've never heard of with Fred's scribbled notes beside them: "Kenya, August 12, 1962" or "Costa Rica, December 5, 1970." Now in his eighties, he is less active internationally but just as busy locally. His eyesight might be failing, but his hearing is as acute as ever, and, as he says, "bird watching is mostly just listening." As we walk along the green pathways of the ravine, I watch him concentrate on the cacophony of birdsong about us as a conductor listens to a symphony, picking out the oboes from the viols, revelling in the deep tones of the alto flutes and riding on the crest of the piccolos.

"Hear that?" he says, making a note on his checklist. "White-throated sparrow. Their call is, 'Dear, Dear, Canada, Canada, Canada.' Unless you're an American, in which case they say, 'Sam, Sam, Peabody, Peabody, Peabody.' I prefer the Canadian rendition myself." Then: "There's a white-*crowned* sparrow, first of the season. They have quite a different call. My old friend Greg Clark used to say their song was, 'Poor Jojo, pissed his pants.'" And he laughs, nearly doubling over at the memory, then straightening and wiping his glasses as we continue along the path: "Oh dear: Poor Jojo, pissed his pants."

This particular day is a fine one from every point of view. As we walk through a very nearly Carolinian tangle of low shrubs and creepers under a thin canopy of newly forming leaves glimmering in the wan spring sunshine, we spot or hear at least two dozen different species of birds, some migratory, brought down by the light rain that had fallen during the night, some full-time Torontonians. Apart from the white-throated and white-crowned sparrows, there are a pair of cardinals, male and female, flickering like fire through the underbrush, a chimney swift (heard but not seen), least flycatchers and great-crested flycatchers, a prothonotary warbler, a whole treeful of rose-breasted grosbeaks, a yellow-shafted flicker, and a bevy of the usual robins, grackles, chickadees, and blue jays. And, of course, house sparrows and house finches. Fred is especially interested in the house finches; they are by no means rare in Toronto these days, but they present an intriguing ornithological puzzle.

"It's a western bird, originally," he says. "When I heard those house finches in *The Savage Innocents*, I knew the film had been shot in California because there were no house finches east of the Rockies in those days. But a few decades ago they began moving east. Then about twenty years ago a small flock was sighted in the Niagara area, and they moved up into southern Ontario and just exploded. There are probably as many house finches in Toronto now as there are house sparrows. Why this should be so is a mystery, since both birds occupy more or less the same ecological niche. Usually, when two species occupy a single niche, one dominates and drives out the other. Does this mean that the house finch is driving the house sparrow out of its traditional territory? And if so, why? What gives it its competitive edge?"

It's a small story, says Fred, as we continue along the path, but it's the kind of puzzle that impelled Darwin to write *The Origin of Species*, and it brings up an array of questions that have to be answered if we are to understand what is happening within our own environment.

Roger Tory Peterson describes the purple finch as "a sparrow dipped in raspberry juice," and the house finch looks so much like the purple finch that it took Merilyn and me several days of lively discussion on the back deck, even with Peterson's guide lying open on the drinks table, before we agreed that the small, brown, red-headed and rose-rumped birds darting in and out of the ivy on our north wall were house finches. In the end it was their song that decided us. If, as Fred renders it, the white-throated sparrow sings a short, repetitive phrase that sounds like "Dear, Dear, Canada, Canada, Canada" (or, as it's given in Nova Scotia, "Poor, Poor, Canada, Canada, Canada"), and the purple finch gives what Peterson calls "a fast, lively warble," the house finch rhymes off a long, intricate, gurgling trill of a sentence that lasts for three full seconds and, whatever byways it takes to get there, invariably winds up with what linguists would call a tag-ending, "chee-ur," which seems to me to be saying "didn't he": "After pissing his pants Jojo took them off and washed them and hung them up to dry and the wind took them out over the meadow but he went and got them back, didn't he."

The house finch (*Carpodacus mexicanus*) belongs to a very large bird family, the Fringillidae (from the Old English *fringil*, meaning "colourful," from which our word "finch" is derived); the family includes such seed-cracking perching birds as grosbeaks, sparrows, and buntings. Among the finches, it most closely resembles the purple finch (*C. purpureus*), differing only slightly in size and coloration. It weighs only twenty grams (a tablespoon of sunflower seeds weighs about fifteen grams) with an overall length of about fourteen centimetres, compared to the purple finch's twenty-five grams and fifteen centimetres. It is sexually dimorphous – the males are the ones with the red plumage on their crown, back, cheek, shoulders, and rump, with a brighter red eyebrow stripe not seen on the purple finch. The red comes from the presence of carotene in their diet – carotene is the product of photosynthesis that makes carrots and pumpkins orange – although the females and juveniles, which eat the same things, are just plain brown except for distinctive, thrush-like streaks on their breasts. This business of the showier male plumage is a putative advantage to the species; at least with the house finch, once the eggs are laid the female does all the incubating while the male does all the food gathering. A bright-red bird sitting on a nest would be an easy target for predators, whereas a dark brown and streaked bird is practically invisible in the nest sites preferred by house finches.

These nest sites are almost exclusively in close proximity to humans, which is why it is called a house finch. Originally a bird of the impact zone between forest and grassland, where coniferous trees provided good nesting sites and indicated a steady supply of water, house finches in the east prefer habitat that most resembles such areas – wide open spaces with a few evergreens, easy access to water – and cities like Toronto are perfectly suited to their tastes: lots of ornamental evergreens, open grassy areas and, at least in our eavestroughs, plenty of standing water for bathing and drinking (on warm days, house finches will drink up to 40 percent of their body weight a day). They seem to like to nest in deep foliage, such as the thick ivy on our wall, or in the centre of cedars and other ornamental evergreens. They are not easily disturbed; one year, a pair of house finches built their nest in a hanging plant on our front porch, not two metres from the door. The city also provides plenty of innovative nesting material. As Ross James, writing in the *Ontario Field Biologist* in 1978, says, "It is difficult to describe a typical house finch nest, as they are placed in a great variety of sites and constructed of a wide range of materials, depending on what is locally available." From her study window on the third floor, Merilyn has

watched a pair of house finches picking the cloth insulation
off the plethora of wires running into our house. We thought
this may have been a macabre way of keeping squirrels from
running along the information highway to get at their nest,
but later that year when a nest fell out of the ivy, I examined it
and found it lined almost entirely with strips of fluffy cloth
insulation.

The house finch's diet consists mostly of vegetable matter:
grass and weed seeds, buds and fruits. At feeders here in the
north they prefer sunflower and niger seeds, apparently for
their oil content, which helps them build up cold tolerance,
but they will also peck at millet and other small, hard, dry
seeds. In the fall, the ivy on our house produces clusters of
small, grape-like fruit that attracts dozens of house finches, so
that the wall under the ivy leaves seems alive with festive
chirping as the birds gorge on the (I suspect fermented)
berries. Unlike other passerines, such as the house sparrow,
house finches do not feed their young on insects.

Fred was right about the house finch's rapid spread throughout eastern North America. The first house finch east of the
Rockies was found at Jones Beach, on Long Island, New York,
on April 11, 1941, by two amateur birders, Richard B. Fischer
and Robert Hines, who promptly reported their find to John J.
Elliott. Elliott, who wrote a weekly bird column in the local
newspaper, the *Nassau Daily Review-Star*, was sceptical but intrigued – it was hard to tell a house finch from a purple finch,
and the house finch had never been seen east of its historical
range on the west coast, stretching from southern British Columbia down to Mexico (it had been introduced in Hawaii,
where it was known as the papaya bird), and reaching inland
about as far as western Wyoming. But he kept his mind and his
eyes open, and the following March was rewarded by finding
seven house finches near a tree nursery in Babylon, on Long
Island, about twenty kilometres northeast of Long Beach –

they were singing from the tops of a group of ornamental evergreens. The next year he went back and saw about a dozen birds; then, in May 1943, he found a nest with four young house finches in it, also in Babylon. The house finch could officially be added to the list of eastern breeding birds.

But how did they get there? The house finch is not a long-distance migrator. Even when they do migrate, they don't gather in huge flocks and suddenly take off, to turn up two days later thousands of kilometres away, as black-poll warblers do. They sort of poke along individually or in pairs, flitting from coppice to fence-row as the food source dictates, until they eventually find themselves in an area where seeds are available all winter. So how, Elliott wondered in his column, did Long Island's growing population of house finches get from Wyoming to the east coast without having been spotted anywhere else in between?

His question was finally answered in 1947 by Edward Fleisher, a doctor living in Brooklyn, who reported to Elliott that in 1940 he had gone into a bird store in downtown Brooklyn and found caged house finches for sale, under the name "Hollywood finches." Fleisher reported this outrage to the National Audubon Society, which, upon investigation, found that nearly every pet shop in New York was selling house finches under one name or another. House finches were classed as migratory birds by the U.S. Bureau of Biological Survey, even though they aren't, and so trapping and selling them without a federal permit was against the law. Most of the shops were supplied by wholesalers in California, which turned out to have sold more than 100,000 house finches to dealers throughout the eastern United States during the 1930s, at the price of $35 per hundred, and getting away with it by calling them either purple finches (which were not protected), "Hollywood finches," or "red-headed linnets," which made them sound like movie stars. They apparently didn't know about papaya birds. When the Bureau of Biological Survey cracked down on this trade in 1940, thanks to Dr. Fleisher, some New York dealers (and perhaps a few of their more knowledgeable customers) quietly took their illicit house finches out onto Long Island and released them.

Ornithologists are not sure how a bird whose natural habitat is the high, semi-arid interior of the west coast managed to survive a winter in the low, damp east, but it did. The speculation is that they survived by keeping to places where the natural eastern habitat has been changed by human endeavour into unnatural western-style habitat: by cutting down the boreal forests and replacing them with grassland and imported evergreens, we created in our cities little microclimates that house finches found congenial. By the late 1940s, the species had spread off Long Island and onto the New York and New Jersey mainland, and since then its increase has been explosive. Coming from Mexico and California, they are not naturally cold-hardy birds, and so their expansion northward has been slow. The first sighting in Canada was recorded in Kingston by Ron Weir in 1970, but it wasn't until 1978 that the first house finch nest in Canada was reported; it was found in a hanging flower pot on the porch of Mr. and Mrs. J. Field, on Front Street in St. Catharines – it was a small, shallow cup-shaped affair supported on a few flower stalks, just like the one we found on our front porch a few years ago. As Fred says, they are now well established in Toronto and are common even as far north as North Bay and Sudbury. Unfortunately, Churchill will never be a good place to look for them; it will have to content itself with being a breeding site for the endangered Ross's gull.

The reason for this sudden cold-hardiness in house finches isn't very well understood, but it seems the species is a highly adaptive one, as if it can't wait to split up and form a new subspecies. House finches from southern California, when introduced into northern Colorado in 1985, took only six months to show improved resistance to cold weather. They did it by increasing the amount of lipid-rich, high-energy food such as sunflower seeds and fruit in their diet, which they are also doing here. The house finches that come to our feeder go for the niger and sunflower seeds, leaving the hard, dry little bulgar seeds for the chickadees, who don't seem to want them either.

This accelerated evolution may explain why house finches have survived so well – there are now an estimated 1.44 billion of them in North America, more or less evenly distributed between the east and the west, with only a narrow strip of house-finch-free treeless grassland running down the centre of the continent. But it doesn't explain why they are taking over the ecological niche formerly occupied by the house sparrow. There is little doubt that they are: in every part of the east that has been studied, house finch increase has coincided with house sparrow decline, until in many areas house finches actually outnumber house sparrows. Why?

Fred Bodsworth thinks the answer has to do with brown-headed cowbirds, two of which flit along the path beside us for a while as we make our way along the ravine. Fred glares at them. They look like monks from some obscure medieval sect, with glossy black robes and dark brown cowls over their heads. The brown-headed cowbird (*Molothrus ater*) is a smallish blackbird, though at forty to fifty grams it is twice as big as the house finch. Cowbirds are related to red-winged blackbirds, grackles, starlings, orioles, and other Icterids. Originally called the buffalo bird, it was found on the Prairies, sitting on the backs of bison waiting for the animal's hooves to stir up its principal food items, mostly weed seeds and, especially during breeding season, grasshoppers and beetles. When we humans eliminated the bison and substituted domestic cattle, the buffalo bird changed its name to cowbird and settled on the backs of cattle instead, and the bird's range increased as fast as we could cut down trees, grow grass, and introduce more cattle. If we've created ideal house finch habitat in the east by altering the landscape to build cities, we created cowbird habitat by changing the landscape to make farms. The brown-headed cowbird may be thought of as a kind of avian coyote.

It arrived in Ontario at the turn of the century and is now found everywhere cattle are, and some places where cattle are not, such as downtown Toronto. Cowbirds don't need cows, they need insects. Toronto is at the southern end of their breeding range, which means sometimes cowbirds will migrate several kilometres south in the fall, and sometimes they won't, depending on the mildness of the winter and the availability of insects. "The annual Christmas Bird Count in Toronto," Fred says, "always turns up a few hundred cowbirds, which means there are likely several thousand of them hanging around."

To me, the fascinating thing about brown-headed cowbirds is their reproductive strategy. They don't build nests of their own, but lay their eggs in the nests of other birds. This behaviour, known as brood parasitism, is not uncommon in the bird kingdom. The European cuckoo is probably the most famous of the brood parasites – the North American yellow-billed cuckoo, though related to the European bird, is not a brood parasite; one female yellow-billed cuckoo may deposit her eggs in another yellow-billed cuckoo nest, which makes a sort of sense (think of it as a form of adoption), but true brood parasites like the brown-headed cowbird go further than that: they lay all their eggs in the nests of other species, which is more like abandoning Romulus and Remus to the wolves. When the female brown-headed cowbird is ready to lay her eggs, usually from late April to early June, she perches at the top of a tree in her home range and watches for nest-building activity in other birds – just about any other birds, although she does have her favourites. Cowbirds have been known to parasitize the nests of 220 bird species, but their favourite hosts – at least in the east – are the various species of warblers, sparrows, phoebes, and vireos.

When the female cowbird finds a host nest, she inspects it for suitability while the host is away – size seems to be important; she prefers nests with inside diameters between ten and nineteen centimetres, and with at least one, preferably two host eggs already in them, as long as the eggs are slightly smaller than her own. Sometimes, if there are two host eggs in the nest, the cowbird will destroy and eat one of them, ejecting the broken shell and replacing the host egg with one of her own. She does this a number of times; studies show that a cowbird produces an average of forty eggs per season, laying one per day, always about ten minutes before sunrise, and rarely depositing two of her own in a single host's nest. She therefore has to find thirty to forty suitable host nests within her home range, a requirement that necessitates careful selection and diligent defending of home territory. This apparently contradictory combination – the intelligent choosing of a stupid reproductive strategy – is highly intriguing: it makes me think I'm misinterpreting at least one side of the equation.

Only 3 percent of all cowbird eggs actually produce chicks that reach adulthood, which is an extremely low percentage and supports the adjective "stupid." Not unexpectedly, some hosts have developed methods of dealing with cowbird invasion. The yellow warbler, for example, perhaps the most frequent victim of cowbird parasitism, simply builds a new nest on top of the old one if it discovers a cowbird egg among its own. Other species, including the eastern kingbird, the robin, the grey catbird, the cedar waxwing, the brown thrasher, and the northern oriole, tip the cowbird egg out of their nests. Blue jays will actually eat the cowbird egg before tipping the shell out of their nest. In fact, if baseball's American League East ever admits a new team, Blue Jays fans should hope it is one called the Brown-Headed Cowbirds.

Effective countermeasures on the part of some birds don't stop the cowbird from continuing to lay eggs in any suitable nest. Red-winged blackbirds, for example, almost always detect and destroy cowbird eggs in their nests, and yet they are, after yellow warblers, the second most commonly parasitized species on the cowbirds' list – probably because they are the most numerous birds in North America. But the point is that cowbirds are particularly unchoosy when it comes to selecting adoptive parents for their offspring. However, the strategy – lay a lot of eggs and hope for the best – works well in many animals (fish, frogs, and most insects, for example) and seems to work well for cowbirds: even at a 3 percent success rate, a pair of cowbirds will produce ten offspring in their lifetime, easily enough to ensure the doubling of the population every four years.

So the balance swings towards "intelligent." The cowbird egg usually hatches ten to thirteen days after being laid, depending on which species is the host. This is a most extraordinary evolutionary adaptation. Different species have different incubation periods, and the cowbird has to ensure that its egg hatches at the same time as those of its chosen host. Hatching too early can be disastrous. The red-winged blackbird, for example, whose eggs hatch in eleven days, routinely cleans its nest of broken eggs for the first nine days and stops on the tenth day, in case it inadvertently tosses out one of its own early hatchlings. If a cowbird egg begins to hatch on the ninth day, it runs the risk of being ejected as a broken egg by the blackbird. So cowbird eggs in red-winged blackbird nests hatch on day eleven. In prairie warbler nests, however, cowbird eggs hatch after 11.8 days; in northern cardinal nests, they hatch in 12 days, and so on. Normally, incubation time is a function of egg size: the smaller the egg, the less time the female has to sit on it. This would seem to imply that cowbirds regulate the incubation time by altering their egg size to suit the host species – as European cuckoos do. But they don't. Cowbird eggs are always the same size; they just hatch, on average, 1.3 days earlier than the time predicted by their size. How they manage to do this is unknown.

European cuckoo hatchlings often mimic host hatchlings in size, coloration, and even early flight patterns, so that host parents may be raising young cuckoos without really being aware of it. Not so with cowbirds. Cowbird hatchlings are always bigger and bolder than the chicks of the host species, so much so they require, and usually get, most of the food brought back by the host parents. Nonetheless, the host parents diligently feed the cowbird chicks, even if it means neglecting their own young. A pair of house sparrows with a cowbird chick in their nest will wear themselves out bringing insects back to the nest, only to have most of the food gobbled up by the cowbird chick, until the adults are nearly dead from exhaustion and the house sparrow chicks have died of starvation. Why the host adults, which must know that that monstrous, gaping maw in their nest does not belong to one of their own kind, continue to feed it to the detriment of their own chicks is another mystery.

Which brings us back to Fred's theory about house sparrow decline and house finch increase. House finch nests are parasitized by cowbirds at least as often as house sparrow nests are – that very first nest discovered on Mr. and Mrs. Field's front porch in St. Catharines in June 1978 had five house finch eggs and one cowbird egg in it. Studies in eastern Ontario show that as many as one house sparrow nest in three contains at least one, and often two, cowbird eggs. Elsewhere the figure can be as high as 70 percent, and the same numbers can be applied to house finch nests. If most of those house sparrow chicks die, the result would be a serious decline in house sparrow numbers.

House finch chicks, on the other hand, would not suffer the same fate, since house finch parents do not feed insects to their young – they bring them weed seeds, sunflower seeds, and other macrobiotic delights. Cowbird chicks do not thrive on seeds, and therefore a cowbird chick in a house finch nest is not likely to hog all the food from its adoptive brothers and sisters, and the house finch population will not suffer the same decline as house sparrows do. The betting is that a good proportion of cowbird eggs in house sparrow nests falls into the 3- percent survival category, whereas all cowbird eggs in house finch nests end up in the other 97 percent.

"It's not the whole answer," says Fred, "because house sparrows are declining in areas where there are no house finches – in some of the southern states, for example. But around here, at any rate, I think the theory has some merit."

After an hour's stroll through the ravine, we turn back, Fred making arcane scratches in his bird book. "Two mourning doves up there sitting in an old squirrel nest," he mutters to himself. "I'll have to keep an eye on that." We see two more house finches near the top of a cedar, trilling away in the spring sunshine. "Finches nest in colonies," he says, "unlike sparrows. I wonder if that helps them fend off cowbirds, too. Maybe they set up some kind of collective warning system." The ravine is beginning to fill with people: parti-coloured joggers, middle-aged women walking dogs, young women in grey business suits and white running shoes hiking to work along the ravine's unkept trails. It is an odd but reassuring mix, showing that the ravine is an accepted part of the city.

Fred's mind seems to have gone back to Churchill and the perfidy of Hollywood filmmakers: "House finches in the Arctic," he says, shaking his head. "They may be moving north, but not that far. It works the other way, too, though. Did you ever see the 1956 version of *Moby Dick*, with John Huston and Gregory Peck? Great film. Ray Bradbury wrote the script. But listen to the birds in the background," he nearly shouts. "European blackbirds! The whole thing, that great American classic, was shot in Ireland!"

SEA-GULLS
IN THE CITY

T he sports incident many of us remember almost as
well as Paul Henderson's goal occurred at Exhibition
Stadium on the evening of August 4, 1983, while the
Blue Jays were playing the New York Yankees. Just before the
Jays came to bat at the bottom of the fifth inning, the Yankees
were warming up on the field. Left-fielder Dave Winfield
tossed one of the warm-up balls to a ball-boy on the sidelines,
and the ball bounced once on the AstroTurf and hit a gull that
was standing on the field, minding its own business, square on
the head. The bird keeled over, the crowd booed and threw
things onto the field, and time was called. A Blue Jay bat-boy
ran out with a towel and took the bird into the Jays' club-
house, where it was examined by the team physician. From
there it was taken to the Toronto Humane Society, where it
was examined again and pronounced dead by the society's
acting general manager, Michael O'Sullivan. The body was
then transferred to an ambulance and driven to Guelph so that
members of the Ontario Veterinary College could perform an
autopsy on it, a procedure that O'Sullivan said was "routine"
when criminal charges were "laid or contemplated."

After the game, which the Yankees won 3-1, Winfield was
arrested by Constable Wayne Hartery, taken down to 14

Division, and formally charged with cruelty to animals – the penalty for which is a $500 fine or six months in jail or both. Constable Hartery told reporters that he had been "upset" by the incident because he was an animal lover and owned a Dobermann pinscher named Runner. Winfield spent the night in jail and was released the following morning on $500 bail, with trial set for the following week. Blue Jays' president Pat Gillick drove down to the police station himself to pick up Winfield, who got into the car shielding his face from photographers with a Blue Jays' game program. Meanwhile, the autopsy on the bird confirmed that it had indeed been killed by a sharp blow to the head from a hard, round object, which only goes to show that diamonds are not necessarily a gull's best friend.

The bird beaned by Winfield was a ring-billed gull, *Larus delawarensis*, one of about 100 that thought they had season's tickets to watch the Blue Jays at the old Exhibition Stadium. They would fly in especially for the games and line up along the third baseline, occasionally waddling stiff-kneed out onto the field to examine objects of particular interest, but generally staying out of the way of the larger white-clad figures that dominated the invitingly green expanse. After the game they would fly out to centre field to look for earthworms, and when they didn't find any they would wheel nonchalantly up into the stands for the bits of hot-dog buns, popcorn, potato chips, french fries, or bratwurst left for them by the day's departed fans.

The return of ring-billed gulls to Toronto is one of the great comeback stories in natural history. In the middle of the last century, the species was so numerous in the Great Lakes region and down the east coast that John James Audubon referred to it as the Common American Gull (it is actually a subspecies of the common gull, *L. canus*, so Audubon may have been lumping commons and ring-bills into one gigantic flock), but human encroachment took a great toll among them. Gulls do not respond well to habitat loss, and city har-bourfronts destroy a lot of gull habitat. Some historians date the decline of gulls in Toronto to the growth of the city's rail-way yards, down at lakeside, in the 1850s. Others remind us that many gull populations were eradicated during the "plume wars" of the late 1800s, when hunters killed entire colonies of nesting birds to supply the millinery trade with feathers, eggs, nests, and sometimes whole birds for the adornment of women's hats. When ornithologist Frank Chapman strolled down New York's Fourteenth Street one day in 1886, field book in hand, he listed forty species of birds whose body parts were perched on the heads of passing ladies: cedar waxwings, flickers, terns, bobwhites, and snow buntings were the most conspicuous. One woman wore the entire wing of a pileated woodpecker, another of a saw-whet owl. Although songbirds were the most popular species taken, the craze soon shifted to shorebirds because their tendency to nest in colonies made them easier to bag in bulk. Cape Cod supplied 40,000 terns a year to New York hatters; a single village on Long Island prepared 70,000 bird skins for the hat trade. Florida became a virtual killing grounds: "It is hard to con-ceive," write Felton Gibbons and Deborah Strom in *Neighbours to the Birds*, a history of American birdwatching, "that this fad once decimated such familiar birds as the herring gull, lesser yellowlegs, egrets and ibises, reducing their numbers to the danger point. Other less common birds like the roseate spoonbill and the curlews may not have fully recovered today from the mass murders of a hundred years ago, and they may never."

The ring-bill did not escape the slaughter. In 1888, the Toronto Humane Society published a tract, "The Wanton Destruction of Birds," citing the plume wars as a particularly egregious example of avian genocide. Under the heading "Birds as a Decoration for Bonnets" was reprinted a doggerel, or rather birdderel, from England's *Punch* magazine, which begins:

> Lo! the seagulls slowly whirling
>   Over all the silver sea,
> Where the white-toothed waves are curling,
>   And the winds are blowing free,
> There's a sound of wild commotion,
>   And the surge is stained with red;
> Blood incarnadines the ocean,
>   Sweeping round old Flamborough Head . . .

and ends with the plea: "There's the riband, silk and jewel, / Fashion's whims are oft absurd; / This is execrably cruel; / Leave the feathers to the bird."

As well as the plume-and-wing trade, the Humane Society joined with the recently formed Audubon Society in railing against all sorts of atrocities, including egg and nest collecting and the shooting of birds for sport. "Not once or twice only at the sea-side," fumed an oratorical clergyman, "have I come across a sad and disgraceful sight – a sight which haunts me still – a number of harmless sea-birds lying defaced and dead upon the sand, their white plumage red with blood, as they had been tossed there, dead or half-dead, their torture and massacre having furnished a day's amusement to heartless and senseless men. Amusement! I say execrable amusement!" And yet bird populations in the area continued to decline. According to W.B. Barrows, author of *Michigan Bird Life*, by 1912 the ring-billed gull had all but "disappeared as a breeding species from the Great Lakes."

By 1960 there were fewer than 60,000 ring-bills in all the Great Lakes, most of them located in Georgian Bay. Toronto itself was almost gulless: a survey of the eastern headland of the outer harbour found only twenty-one nests in 1973. But that was the last year anyone had to be concerned about the ring-billed gull population. By 1976, the same eastern head-land (better known as Leslie Street Spit) was the site of 10,000 nests; by 1984, there were 74,500 nests — nearly half the nests found on Lake Ontario and more than 12 percent of all nests on the Great Lakes. The ring-billed gull had regained its former status as a "problem species." Clearly, Dave Win-field's bean-ball notwithstanding, Toronto's ring-bills have staged a spectacular drive in the late innings and are threaten-ing to sweep the series. On a recent trip to the Leslie Street Spit, I visited the eastern headland gullery — one of the largest ring-billed gull colonies in the world — and was amazed at the number of gulls nesting on the spit's sandy isthmus: the point was white with gulls, the air was deafening with their cries. I calculated two nesting pairs every square metre, with as many others walking or standing about and still others hovering in a holding pattern above as if waiting for a space to land. Say what we like about gulls, they seem to like it here, so much so that they have changed their entire way of living in order to spend more time with us, sharing our beaches and our parks. So we might as well get used to them.

Ring-billed gulls are a migratory species. Those that spend their summers on Lake Ontario begin to congregate in late August and early September, and when food supplies begin to dwindle in October they head east in small, loosely associ-ated flocks of six to three hundred birds, flying low across Pennsylvania and New York State to the Atlantic coast, stop-ping to forage for food in farmers' fields and irrigation ditches, or the mall parking lots and landfill sites that have replaced them, and roosting at night near fresh water. They sort of drift south, like bush-league ball players doing a series of small-town tournaments, not in any great hurry, flying low, circling back, travelling maybe 100 or 200 kilometres a day, staying in one place just as long as the food holds out. They work their way through North and South Carolina and Geor-gia until they hit the east coast of Florida, where most of them spend the winter fattening up for the return flight. Some of them circle around to the Gulf coast and the lower Mississippi Valley. Then in mid-February they head back north, following a slightly different route (they veer west at Chesapeake Bay to link up with the great Mississippi flyway; no one knows why) and travelling at a more determined pace, seldom stopping at all. By mid-March they are back on Lake Ontario, staking their claims to the half-square metre of Leslie Street Spit that is their northern home.

The first thing the female does when she arrives here is start rebuilding her energy reserves in preparation for breeding. This is when winter-killed fish come in handy, as well as spawning salmon and rainbow trout, landfill sites and spring ploughing. Meanwhile, the male builds the nest (out of dead plant material), or repairs the one it used the previous year, and occupies it until the female comes to him begging for food, a sure sign that she is ready to mate. This act is accomplished with much of what bird biologists call Wing-flagging and Copulation-calling, with the male standing on the middle of the female's back while she Head-tosses and strokes his chest feathers with her beak. He then slides off as she raises her tail and there ensues much Tail-wagging while cloacal contact is made. Ring-billed gulls were once thought to mate for life — but then scientists of the moral majority once thought all animals, even mosquitoes, mated for life, or ought to. Recent studies suggest that two or three seasons is the average length of gull marriages: grounds for divorce are usually incompatibility with regards to egg incubation. The eggs are laid during the last ten days in April, and both males and females share the task of sitting on them about equally (both have brood patches on their chests, areas of bare skin that allow direct contact of warm skin on the egg), and both seem to derive great pleasure from it. Squabbles over whose turn it is to sit on the nest can go on only so long before the lawyers are called in.

Western ring-bills have often been observed kicking egg-sized pebbles into their nests and sitting on them as though they were eggs, and there is some debate among ornithologists as to why they might be doing this. The average clutch size for ring-bills is three eggs; gulls that seem to have developed a liking for faux eggs will lay one or two eggs and then fill out the clutch with one or two pebbles. When researchers remove the pebbles from the nests, the gulls roll them back in again. When the real eggs are taken out and only the pseudo-eggs are left, the gulls continue incubating the pseudo-eggs. Some researchers hypothesize that the pebbles function as incubation stimuli, that gulls with smaller-than-average clutches are actually placing the pebbles in their nests in order to make themselves think they have a full complement, and they will therefore tend them more assiduously. Others suggest that the gulls had originally mistaken the pebbles for food, swallowed them, regurgitated them near the nest, then said to themselves, "Gee, that wasn't a peanut-butter sandwich after all, that was one of my eggs," and toed the thing back into their nest. Closer study, however, reveals that gulls rarely mistake pebbles for food, whereas the incidence of pseudo-eggs in the nests of certain colonies is quite high; also, most of the pseudo-eggs come from very close to the nests, and it is unlikely that a gull would sit on its nest for several days staring at an egg-sized pebble, and then suddenly decide that the pebble was in fact a peanut-butter sandwich. Much more likely that the brooding gull would mistake the pebble for one of its own eggs, which is not necessarily a mark of stupidity. Many birds show a remarkable inability to distinguish their own eggs from the eggs of other species — hence the enormous success of such nest-parasitizers as cuckoos and brown-headed cowbirds. But some birds, including ring-billed gulls, seem to have difficulty distinguishing their own eggs from inanimate objects — Canada geese, for example, have been found incubating pine cones. Ring-bills incubate pebbles because they think the pebbles are eggs – or at least they suspect the pebbles might be eggs and decide that it makes more reproductive sense to be safe than sorry. It doesn't though, because if it weren't for the pseudo-eggs, the gull might go on to lay one or two more real eggs, thereby increasing its chances of reproductive success by a factor of two or three.

If loss of nesting habitat caused by human encroachment accounted for the gulls' decline at the beginning of this century, creation of nesting habitat by human activity is in large part responsible for their subsequent increase. Viewed from the point of view of a gull, the Leslie Street Spit is heaven: gulls like to nest on low ground, in close proximity to water, in sandy soil with a variety of sparse or willowy vegetation. This is exactly the kind of habitat that was destroyed along the lakeshore in Toronto from about 1850 onwards, and it is exactly the kind that is being created now by our landfill activities, not only on the Spit, but along the whole Lake Ontario shoreline, from Bluffer's Park in Scarborough to the Humber estuary.

A list of the food preferred by ring-billed gulls is still topped by fish, but it's a long list, and one that has changed considerably over the past two decades as gulls have become more dependent on land-based food sources. One of the reasons for the gulls' amazing come-back, apart from the cessation of the plume wars and improved habitat, has been the introduction of new fish species into Lake Ontario. The alewife (*Alosa pseudoharengus*), for example, a saltwater fish that was accidentally introduced into the Great Lakes in 1873, has become one of the most abundant small fishes in Lake Ontario and is a staple diet of most shorebirds, including the ring-billed gull. Notoriously incapable of withstanding changes in water temperature, the alewife population fluctuates wildly with the weather – the severe winter of 1993–94, for example, just about wiped out the entire alewife population in Lake Ontario – and vast numbers of them die and are washed up on shore, to the disgust of cottagers and the great delight of scavenging gulls. The alewife is a member of the herring family – in the Atlantic provinces it is known as the gaspereau – and is a favourite food of the herring gull. For a long time the ring-bill and the herring gull competed for it, with the larger herring gull usually winning. But any battle between two species for food is energy-consuming to both species, and so herring and ring-billed gulls have come to a kind of unnegotiated truce; they have mutually decided to develop tastes for less disputed items, especially during the breeding season. Both species dine together on dead alewives about half the time and then split off to forage for different and non-overlapping food items the other half of the time.

Herring gulls and ring-bills went inland for their second course, but they went after different foods. A study conducted in 1976 of herring gulls in eastern Lake Ontario found that their diet during the breeding season, from mid-April to June, consisted of 50 percent alewives and 50 percent small mammals (mostly meadow voles, *Microtus pennsylvanicus*) and, at least for the first two weeks of May, migrating birds – species identifiable in herring-gull spew included ruby-crowned kinglets, red-winged blackbirds, blue jays and sparrows, but there were undoubtedly many more. The gulls, it seems, would wait along the north shore of the lake for flocks of long-distance migrants to struggle over the wide expanse of Lake Ontario and then pick them off as they slumped exhausted and starving to the ground in Canada.

Ring-billed gulls are no more sporting than herring gulls and are not above scavenging dead and dying passerines when they have the chance. Every night in early May, hundreds of migrating songbirds flying over Toronto bash themselves into the city's downtown office towers, attracted by the towers' bright lights and confused by all that glass – and fall stunned to the ground. And every morning in early May hundreds of ring-billed gulls pad along Bay Street, scooping them up for breakfast. May is also gull-breeding season, and to them a stunned or dead warbler is instant protein. In April 1990, wildlife artist Michael Mesure began collecting the wounded birds in order to rehabilitate them and return them to the air, and on one of his early-morning excursions he ran into Carolynn Parke, who worked in a law office in the Royal Trust Tower and had been rescuing birds on her own for a year. Realizing that more than two concerned souls might be willing to brave the chilly May mornings to rescue dazed birds, Michael and Carolynn organized the group they call the Fatal Light Awareness Program (FLAP). Pretty soon there were about ten members, and they began patrolling the sidewalks around Bay and King every morning from late April to the end of the spring migration, about the third week in May, and again during the fall migration, in September and October. In 1993, their first year of operation, they collected 1,292 birds, representing a total of seventy-one species.

One cold morning in May, I turned up at the base of First Canadian Place at five o'clock to lend a hand. It was still dark, and the only traffic on the streets were a few taxis and a street-cleaner. There were five of us; I made the rounds with Irene Fedun, a compact, athletic woman who worked in a print shop during the day and edited *Touching Down*, FLAP's newsletter, by night. I was holding a cup of Mars coffee; Irene was carrying her FLAP equipment: a flashlight, a tape recorder, a small net, and a backpack full of brown-paper lunch bags. Whenever she found a wounded bird, she would pick it up, warm it in her hands for a few minutes while recording the species, date, time, and location into her tape recorder, and then put the bird in a paper bag until she could pass it on to Wendy Hunter, who works in the Wildlife Department of the Toronto Humane Society and is also a founding member of FLAP. Later, Wendy released the birds in High Park or some other semi-natural spot from which they could resume their migratory flight.

Our first find was an ovenbird (*Seiurus aurocapillus*), a tiny, buff-coloured wood warbler, lying among some leafless bushes near the southern entrance to the TD Centre. It was dead. "We get a lot of these," Irene said, putting it in a bag. Dead birds go to the Royal Ontario Museum. "They're very delicate, easily panicked, and they usually don't survive the crash." Of the 226 ovenbirds the group found last year, 133 were dead. The most common species rescued by FLAP were white-throated sparrows (*Zonotrichia albicollis*), at 258 — larger and somewhat hardier birds, but still vulnerable. "White-throats are in serious decline in the wild," Irene said as we circled the Royal Trust Tower, her flashlight scouring the planters in the wind-whipped canyons. "We don't really know why; maybe it has something to do with their predilection for crashing into lit office towers." We picked up two more birds coming up Bay: a dead hermit thrush (*Catharus guttatus*) and a live wood thrush (*Hylocichla mustelina*), both of which Irene duly recorded and put into bags.

At sun-up, we reunited at King and Bay to compare our finds. There weren't many; it had been a slow morning, too early in May for the full migratory flocks. We had three more white-throats, a golden-crowned kinglet, and a Magnolia warbler. The five of us went for a quick coffee, and on the way, Ron Tomus told me that one of FLAP's more important mandates was to convince the owners of these skyscrapers to turn off their lights at night, a battle energy conservationists had given up on a decade ago. FLAP was making some headway, Ron said. Michael Mesure was having encouraging meetings with Cadillac-Fairview, and the group could already claim credit for one significant change to the Toronto skyline: "Birds are most attracted to white light," Ron said, "then yellow. See the symbol at the top of the Canada Trust building?" And he pointed up to the huge, square inverted horseshoe that is the Canada Trust logo. It was red. "Last year it was yellow," said Ron. "We got them to change it because red is less attractive to birds. It's a small change, but it's a start."

As we went into a coffee shop near Bay and Front, the sun was just beginning to warm the pavement, and the first gulls were flying in to see if we'd missed anything.

Ring-bills go farther inland than the TD Centre in search of non-traditional meals, and for most of the summer the food items they choose are vastly different from the field mice sought by herring gulls. They go for earthworms and insects. A study conducted in the 1960s of ring-bills in Michigan found that fish (mostly alewives, but also some yellow perch and nine-spined sticklebacks) provided 32 percent of the gulls' diet, but the rest was made up of worms, mayflies, damselflies, beetles, cicadas, midges, and ants. Earthworms become the gulls' principal food item in April and May, when farmers do their spring ploughing, with fish and insects coming in second and third respectively. In June and July, however, insects predominate, forming up to 89 percent of the total volume. The authors of the study suggested that "the reliance of ring-billed gulls on insects as an energy source is a recent innovation" that "may have reduced the dietary overlap between ring-billed and herring gulls and contributed to population changes."

Those of us who have spent any time at all eating along Toronto's harbourfront in recent years will have noticed that the above list contains no hot-dogs, french fries, Twinkies, bread crusts, KFC bones, or macaroni salad, and this will seem to us to be an oversight. It is true that gulls will eat just about anything tossed at them (and who are we to criticize them for that?). They are scavengers, after all – even the fish they eat is tossed to them by the wind and waves. Their propensity for uncritical acceptance of food has given a word to the English language: in 1597, the Elizabethan playwright Thomas Nashe applied the term "gull" to any "slow, yce-brained, beef-witted" dupe who would believe anything he was told, no matter how absurd – the source of our word "gullible." Thomas Dekker wrote *The Gull's Hornbook* in 1609 as a guide to gulling, i.e., making an ass of someone, maligning a different species. Actually, the award for Most Species Maligned in a Single Sentence goes to Sir Toby Belch, who, in Shakespeare's *Twelfth Night*, asks Sir Andrew Aguecheek if he would deign to "help an ass-head, and a coxcombe, and a knave: a thin-face'd knave, a gull?"

At any rate, during the non-breeding season, when high-energy food is not so important, ring-bills are not what one might call selective feeders; they will pig out on the same nutritionless stuff we pig out on. Some city gulls go for fast food even during breeding season: ring-billed gulls in Bedford, Ohio, for example, do most of their food gathering at the city dump, even during the spring mating season – they eat some fish, a few insects, and the odd earthworm, but the bulk of their diet consists of fried chicken and ribs. Bedford gulls, admittedly, have no easy access to a lake, as do Toronto birds, but their shift to a predominantly take-out diet is a classic example of adaptability. A woman I know from Stockholm, Sweden, tells me that in her city there are gulls that have seemingly lost their ability to catch fish: they swoop down on unsuspecting humans and steal the hot-dogs right out of their hands. Stockholmers call them *korv mås*, or "sausage gulls." Male gulls do not become sexually mature for two and a half years and therefore do not have to worry overmuch about keeping their diet up to reproductive standards; most of Toronto's sausage gulls along the quays and in the parks are indeed first- and second-year gullants killing time at their version of the local mall.

Gulls are the quintessential urban animal. In Ethel Wilson's brilliant short story, "Sea-Gulls in the City" (from which I have stolen the title), gulls strut proudly in a Vancouver park: "They stand still. Then with the air of one taking up a collection they walk away majestically." The gulls, Wilson writes, "know on which side their bread is buttered," and that is undoubtedly true. They are reasonably intelligent, highly adaptable, sociable, and classically unperturbed by unfamiliar frenzied activity. A bus can bear down on a gull without giving it a heart attack. I've seen gulls riding uptown on the roofs of trolley cars. Once when I was at the Toronto Wildlife Centre, a woman brought in a wire cage covered with a green-and-orange beach towel. Inside the darkened cage, gazing calmly at the beach scene, was a large ring-billed gull. "I found it on the corner of Dufferin and Bloor," said the woman. "It was just walking around, looking sort of dazed. I thought it was going to be hit by a taxi or something." There was nothing wrong with the gull; gulls spend a good part of their waking hours just walking or standing around, looking sort of dazed – about 34 percent of their time, in fact. The woman said she chased the gull for half a block, trying to get it into the cage: "Isn't that strange behaviour?" she said, meaning the gull's. With two schools, a doughnut shop, a mall, and a subway station nearby, Dufferin and Bloor is a busy corner. I thought the gull was probably hanging around trying to figure out what all the *people* were doing there. Or maybe it just wanted to stay close to the doughnut shop.

Gulls in the city pose the same kind of problems that pigeons and Canada geese do. They defecate a lot, and after a while people start talking about health risks. I don't think we really worry about health risks, I think we are simply put off by the sight of so much guano on our well-scrubbed city sidewalks and parks. Strolling along some of Toronto's lakeshore pathways can be a pretty dicey thing to do. The word slipshod comes to mind. In the mid-1980s – around the time of Winfield's wallop – Toronto's beaches were closed for several summers because of high counts of *Escherichia coli* in Lake Ontario's warm near-shore waters. Gulls were blamed for the closings, even though studies showed that the bacteria were coming from somewhere else – probably from our own inadequate raw-sewage–disposal system. But people tend to remember the charges, not the acquittals.

There are health risks associated with gull droppings, and we might as well be aware of them. Aspergillosis, a lung disease caused by inhaling the spores of the fungus *Aspergillus fumigatus*, is found in many domestic and wild fowl and has been reported in ring-bills in the Toronto area; so far it has not been detected in human beings in this area, although there is no clear reason why not. Histoplasmosis is another fungal lung disease that is more often associated with bat and pigeon dung than with gull guano, but there have been two outbreaks in the Great Lakes area, and both were attributed to ring-billed gulls from the gullery near Rogers City, Michigan. Since gulls are highly mobile – Toronto's gull colonies have been augmented by gulls from nineteen other gulleries on every Great Lake except Superior – there is a chance that some of the Rogers City gulls are now resident in Toronto. But histoplasmosis is usually contracted by people actually working in guano-infested areas – people cleaning out bat caves (for some reason), or tearing down barns or sheds that have been long-time pigeon roosts – and anyone engaging in that sort of activity ought to wear a mask. The two cases in Rogers City involved people working in close proximity to the gullery: landfill workers cleaning a guano-coated bulldozer, and several graduate students studying the behaviour of ring-billed gulls.

Two bacterial diseases – botulism and salmonellosis – are found in gulls. Botulism is most prevalent in late summer and early fall, when water levels are low and water temperature is high. According to a report published in 1986 by the Canadian Wildlife Service, between 1959 and 1964 more than 12,000 waterfowl died in the Toronto area from botulism, but since then outbreaks have been fewer and more confined. Salmonellosis has become more prevalent in ring-billed gulls since they started eating take-out chicken at landfill sites, understandably enough, but so far no human beings have caught it from gulls.

More worrisome than disease, at least to farmers, is the gulls' newly developed taste for earthworms. Earthworms are an essential component in the ecology of soil; they exist to turn plants into dirt. They do this by taking dead leaves in at one end and excreting them through the other end in the form of castings. They are living composters, which is why some of us keep trays of them in our basements and feed them our kitchen scraps. They also play an important role in soil distribution; some species take leaf mould in at some depth, then come up to deposit their castings on the surface, bringing up dirt from as much as a metre down and keeping the whole mixture churned and aerated, a highly beneficial occupation. Losing significant numbers of earthworms to flocks of marauding sea-gulls would naturally lead farmers to look for consequent drops in soil nutrition and crop yields. And farmers are always finding those.

But it is unlikely that ring-billed gulls following a tractor for a few weeks in the spring are going to make deep inroads in the earthworm population. There are a lot of earthworms out there. Charles Darwin, whose last book was about earthworms, reports and agrees with German estimates of 53,767 earthworms per acre of farmland (such estimates tend to be fairly constant across the temperate zones). More recent Swiss calculations put earthworm biomass at up to 2,000 kilograms per hectare. Any way you count them, that's a lot of earthworms. An individual gull will eat about 150 grams of worms per day: 1,000 gulls could clean out a hectare in two weeks, or 10,000 gulls could do it in two days, but few farmers outside Daphne du Maurier's *The Birds* (it was Hitchcock who turned them into crows) are visited by that many gulls. And anyway, as soon as the gulls left, new earthworms would move in from neighbouring hectares, and life would go on.

People with coffee tins taped to their legs and flashlights strapped to their hats, who comb the city's parks and lawns at night for "nightcrawlers" to supply American bait shops, probably account for the disappearance of more earthworms per year than all the ring-billed gulls on Leslie Street Spit.

As a result of these and other supposed gull-related problems (tomato farmers say gulls account for $50,000 a year in lost revenue, and officials at Pearson International Airport complain about gulls congregating on the runways, posing a serious threat to aircraft landing and taking off), during the 1980s – when someone estimated there were 160,000 gulls nesting on Leslie Street Spit, a number that was thought to be increasing at the rate of 10.9 percent a year – there were several calls from Toronto alderpersons to reduce the population to a manageable 10,000 or so, and its growth rate to zero. One plan, championed by Councillor Tony O'Donohue, was to spray the gulls' eggs with kerosene, an oily substance that would clog up the eggs' pores and cause the embryos to suffocate. The Canadian Wildlife Service nixed that idea, pointing out to Tony that not only was it inhumane, but it wouldn't work: more gulls would just wheel in from the other 170 colonies around the Great Lakes, and there wasn't enough kerosene in all of Christendom to keep them at bay. Then a certain Dr. Dmytro Buchnea, a retired University of Toronto biochemist, suggested that gull eggs could be harvested from the Spit and sold as a delicacy to Toronto's restaurants, "as is done in some European countries." He said he could hire students to do the harvesting. The plan even received the approval of the Metro Water Pollution Committee, which said it was "listening to everything that comes along," but bogged down when it was pointed out, again by the CWS, that gulls are in many respects a lot like chickens; they are programmed to lay a certain number of eggs per breeding season, and when you remove eggs from their nest they just keep laying more until they find themselves sitting on their biological quota of eggs. Even if the students replaced each stolen egg with an egg-sized pebble, a task too tedious to contemplate, Toronto's restaurateurs would still have opined that although gulls' eggs are indeed sold in some European restaurants, they are considered déclassé substitutes for the rarer and (reputedly) more delicious plovers' eggs. The CWS had only to remind Dr. Buchnea that ring-billed gulls are a protected species, and any attempt to tamper with them would liable the tamperer to an even more severe penalty than the one the City of Toronto tried to level on Dave Winfield.

In the end, the city opted for a live-and-let-live policy as far as the gulls on the Spit were concerned. They installed a few electronic "shell-crackers" to frighten gulls off their nests during egg-laying season; they hired a hawker to tether a few falcons and ferruginous hawks on the headland from time to time (gulls are deathly afraid of hawks and will abandon their nests and feeding sites when a hawk's shadow passes over them). But the gull problem envisioned by city councillors in the 1980s hasn't really come to pass. The population on the Spit, according to CWS's gull specialist, Hans Blokpoel, has settled down naturally to about 50,000. "They saturated their nesting area," Blokpoel says. After overpopulating their own sites (at one point, there were as many as four nests per square metre), they invaded the nesting territories of their neighbours, the common terns. CWS worked with the Metro Toronto Conservation Authority to provide new nesting sites for the terns – those floating rafts you can see just off-shore of the Spit are the result – but the gulls continued to increase in number. "When they did that, they could either nest even more densely or they could move to other nesting sites," says Blokpoel. Eventually, the surplus Spit gulls chose the latter. They've taken up residence in places like Bluffer's Park, or the Lakeview Generating Station in Mississauga, maintaining their relationship with the lake, but firming up their contacts with the city. A few of them – all right, maybe a few thousand

of them – have moved inland a kilometre or so and taken to
nesting on the flat roofs of commercial buildings along Highway 427, between the Gardiner Expressway and the airport,
much to the consternation of the buildings' owners, who
claim the birds are befouling their buildings' downspouts with
their droppings and causing a lot of roof-top flooding. The
birds could be doing it on purpose; after all, they do still like
to be near water.

Ring-billed gulls are almost instantaneously adaptive.
Blokpoel likes to illustrate this point with the story about the
time, in the spring of 1992, when the lake level rose particularly high, and a large part of the gulls' forty-one hectares of
nesting area was under water. Many of the birds left to find
more suitable habitat somewhere else, but a few of them – just
five or six pairs – took to nesting in the trees that were growing along the headland's western shoreline. "They took up
residence in some old cormorant nests from the year before,"
Blokpoel says. "We climbed up and found gull eggs in them.
Imagine! Of course, the cormorants came back a week or two
later and kicked the gulls out," he adds, but it does show just
how easily a few million years of evolution can be erased.

Nesting in trees, dining on earthworms, blocking downspouts; it doesn't sound much like traditional gull behaviour.
I've seen gulls in their natural habitat – out at sea, sometimes
weeks out at sea. I've seen kittiwakes, wilder members of the
gull family, at the North Pole – and they looked so perfectly
at ease, so unquestionably in control of their own destinies,
that the idea of their ring-billed relations immigrating inland
seems like a betrayal, like a Viking telling his mates that he
wanted to settle down in Brittany and take up goat farming
rather than go back to sea. But even in the city, gulls are a
haunting echo of their wilder kin. As Ethel Wilson reminds
us: "Something shakes for an instant the calm of a man crossing the street when he hears the cry of a gull above the traffic,
something that is not a sound but a disturbing, forgotten,

unnamed desire, a memory. Java, Dubrovnik, the Hebrides. What is it?" Perhaps it is our own wilder selves.

It also helps to remember that successful adaptation, if it goes on long enough, is called evolution, and the next time you see a gull mashing down a Twinkie or playing chicken with a tractor, try to remind yourself that what you are really watching is evolution in action. When you watch a flock of ring-billed gulls pecking at earthworms in a park, you are observing a shorebird turning itself into a land bird: you are watching a gull becoming a robin. Eventually, only those gulls with characteristics more suited to walking on land than to floating on water, or with beaks better built for pecking than for scooping, will survive to pass their genes on to their offspring, who will pass their genes on, until one day no gulls will have webbed feet, and they will not like the taste of fish. They will not make nests on gravel, and so they may not be predominantly grey and white. They will no longer be gulls.

DAYS OF
WHINE AND ROSES

When Allied soldiers liberated Naples in 1943, they were greeted with "delirious happiness" by the Neapolitan citizenry, according to the American journalists who had been following the Italian campaign. The retreating Germans had left behind a bombed-out ruin of a city and a half-starved, typhus-infested populace, giving new meaning to the old Italian saying, "See Naples and die." One of the first things the Allies did was line the entire population up and spray them with a brand-new de-lousing powder that was then being manufactured in the United States. It had been known since the turn of the century that body lice were the carriers of typhus fever, and typhus had killed more soldiers on the Balkan and Russian fronts during World War I than bullets had. The Allies wanted to rid the Neapolitans of body lice so they wouldn't pass on typhus to the troops.

The pesticide they used was called dichloro-diphenyl-trichloroethane, or, as it was shortened in the stenciled inscriptions on the containers, DDT. DDT was a chlorinated compound that had first been synthesized experimentally in Germany in 1874, but it wasn't until 1939 that Swiss chemist Paul Müller discovered that it was a highly effective insecticide, working on the insect's nervous system and killing it in its larval stage. The Americans dusted the Neapolitans with one pound of DDT for every fifteen persons; the population of Naples at the time was almost half a million. It took them several months, but the lice were eradicated and the typhus epidemic died out that winter. No one got sick from the DDT. The Allies went on to win the war.

You can't help but think the dusting of Naples was regarded as a huge experiment, at least in retrospect. Someone somewhere must have said (we'll assume after the war), that DDT was safe, because no one in Naples suffered from the treatment. However it happened, no DDT was used in the United States before 1944, but thereafter it became the insecticide of choice almost instantaneously. DDT powder was mixed with bunker oil and sprayed on crops in the United States to kill the seemingly endless array of insect pests that attacked coniferous forests and food crops such as potatoes, tomatoes, apples, and grain. It was also used against nuisance bugs such as mosquitoes, black flies, and fire ants. Proponents of better living through chemistry began to speak of a brave new world, free of all dangerous or even vaguely unpleasant forms of life, a world released from the cold determinism of unchecked nature.

But there was a fly in the ointment. In powder form, DDT was relatively harmless to humans because it did not penetrate the skin, but when mixed with oil the compound invaded body tissue and became fat-soluble. Even before 1948, the year Paul Müller was awarded the Nobel Prize in medicine for his work on DDT, reports began to appear in various U.S. and Canadian health department offices warning about the toxic effects of the ubiquitous insecticide. Studies showed that, a mere five years after its introduction, virtually no agricultural food product in North America was DDT-free. DDT sprayed on lakes and rivers killed fish. DDT sprayed on meadows killed bees (but not before they had produced DDT-laced honey). It even defoliated plants. The Department of

Agriculture set a maximum of 7 parts per million (ppm) as a safe limit, but some foods sold in grocery stores were already measuring two or three times that level. Medical studies showed, moreover, that because DDT was fat-soluble, it was stored and accumulated in the body, so that a steady diet containing as little as a tenth of 1 ppm soon became contamination on the order of 15 to 20 ppm. They also showed that DDT destroyed essential enzymes in heart muscle, caused severe disintegration of liver tissue, and was carcinogenic.

By the time Rachel Carson's *Silent Spring* was published in 1962, in which she warned of the long-term effects of DDT on human and environmental health, spraying DDT was so common that tocsins started going off from the grass roots right up to the U.S. Senate. Carson spoke eloquently against the use of DDT, pointing out that the chemical was so toxic and non-specific that it was killing not only target insects, but also beneficial insects (like bees), as well as animals higher up the food chain such as fish and songbirds. A 1947 study conducted by the Ontario Department of Lands and Forests reported "considerable mortality among amphibians, reptiles and some species of aquatic life" in a coniferous forest after a DDT campaign against spruce budworm. Bald eagles and peregrine falcons became endangered species because DDT caused them to lay eggs with shells so thin that they did not hatch. Carson also showed that many target insects, including several species of mosquitoes, were developing immunities to DDT, which was leading to the development of ever more dangerous chemicals.

Carson called the chapter in which this possibility was explored "The Rumblings of an Avalanche" and quoted Canadian zoologist A.W.A. Brown of Guelph University, who said that "barely a decade after the introduction of the potent synthetic insecticides in public health programs, the main technical problem is the development of resistance to them by the insects they formerly controlled." Two years after the de-lousing of Naples, she wrote, DDT was used to control mosquitoes in the United States and Canada, and a year later, "both house-flies and mosquitoes of the genus *Culex* began to show resistance to the sprays. In 1948 a new chemical, chlordane, was tried as a supplement to DDT. This time good control was obtained for two years, but by August of 1950 chlordane-resistant flies appeared, and by the end of that year all of the houseflies as well as the *Culex* mosquitoes seemed to be resistant to chlordane."

Resistance to Carson's warnings came from many quarters, including some that ought to have rallied to her defence. A.W.A. Brown, for instance, had read Carson's "Avalanche" chapter in proof and had suggested some changes, which she made, but he still told *Globe and Mail* reporters in 1962 that Carson had wildly overstated the dangers of insect resistance to DDT, saying the problem was more akin to the movement of a glacier than to that of an avalanche and blaming Carson for being the tool of certain "individuals concerned with wildlife work, having a vested interest in opposing pesticides," when she ought to have been listening to "competent public servants" like himself.

Another of those competent public servants was C.H.D. Clarke, then chief of the Fish and Wildlife branch of the Ontario Department of Lands and Forests. Clarke sided with Carson and wrote a strong rebuke to Brown, saying that "every intelligent and sensitive person in the world has a vested interest in wildlife; and that we have a responsibility for the interest of those not yet born."

Largely as a result of investigations urged by Carson, DDT was eventually banned in the United States and Canada, but only after decades of struggle. As Frank Graham noted in *Since Silent Spring*, a 1970 survey of responses to Carson's polemic, "*Silent Spring* was the beginning of that crusade which persuaded administrators and legislators alike that the chemical industry would not act in the public interest unless forced to by stricter regulations." By 1970, the use of DDT had been restricted in Michigan, Arizona, California, England, Sweden, Denmark, Australia, and Hungary. In Ontario, legislation against DDT was passed on September 24, 1969, and went into effect January 1, 1970.

But legislation doesn't automatically do away with a problem. In some ways, it can deflect attention away from an important issue, because people assume that the government is dealing with it — we trust governments when they seem to be doing something we want them to do. In 1985, the breast milk of a woman in Toronto was found to contain 40 ppm of DDT. The woman had emigrated from India, and a chemist at the Ontario Ministry of Agriculture said that "DDT has been disallowed in Canada since 1970, but is still used in India." This sounded too convenient to me, so I looked up the legislation. The Memorandum to the Ontario Pesticides Act, passed in 1970, did not "disallow" the use of DDT; it merely prohibited its use in buildings or pastures used by animals whose milk or meat was intended for human consumption. A dairy farmer could not use DDT in his barn to keep houseflies down. But truck farmers could continue to use DDT in their potato and tomato fields as long as they stopped doing so three to six weeks before harvest, depending on the crop. The Department of Lands and Forests discontinued its use of DDT on Crown lands, but private individuals could continue to use it. The legislation stipulated only that it could not be used "in such a manner that the substance comes in contact with an area other than the area to be treated," whatever that meant. DDT was specifically okayed for mosquito, black-fly, and sand-fly control, as long as its use was intended for "adult control by ground application under provincial permit." It could not be used on cats.

Just how much DDT was used in Ontario after 1970 is unknown, probably unknowable. Resort and hunt-camp owners, the forest-products industry, suburban developers, anyone with an interest in short-term mosquito control could legally use DDT until its sale and use were finally made illegal in 1990. And although DDT probably has not been used extensively anywhere in North America since then, it is still one of the eleven most common pesticides found in Canada from the Great Lake waters to the Arctic Ocean. Because DDT has a half-life in soil of fifteen years, even the huge amounts of it used before 1970 will continue to contaminate our farmland until the year 2004.

Despite all this, DDT hasn't made much of a dent in the mosquito population. There are at least as many of them today as there were 160 years ago, when Anna Brownell Jameson called them "as pretty and perfect a plague as the most ingenious amateur sinner-tormentor ever devised." Jameson was returning to Toronto after an extended boating excursion up the Great Lakes, and mosquitoes were her constant lament. She was taken in an open boat through clouds of them so dense she could not see her oarsmen, let alone the shore. Ever the naturalist, she went on to describe in some detail the insect she was busy swatting. "Observe," she continued, "that a mosquito does not sting like a wasp or a gad-fly; he has a long proboscis like an awl, with which he bores your veins and pumps the life-blood out of you, leaving venom and fever behind. Enough of mosquitoes – I will never again do more than allude to mosquitoes; only they are enough to make Philosophy go hang herself, and Patience swear like a Turk or a trooper."

Jameson was right about mosquito anatomy. We talk about mosquito "bites," but in fact mosquitoes have no biting parts, as black flies do, but rather a complicated and highly evolved (they have been perfecting it for more than 200 million years) system of tubes and pumps designed to extract blood from their victims with the minimum amount of effort. The system comprises a long, piercing proboscis enclosed in a sheath; inside this sheath are six stylets, four of which (two mandibles and two maxillae) are cutting tools equipped with fine teeth, and the remaining two, when pressed together, form a hollow tube. When the proboscis is awled into a host, a secretion of the mosquito's salivary gland flows down the proboscis into the host's blood — this anticoagulant thins the blood so that it can be readily sucked up through the tube. There is actually a small suction pump in the mosquito's head that draws the blood up the tube and into the abdomen: if you watch a mosquito swelling up with your blood, it looks as though it's taking a lot, but in fact each blood meal takes about one-millionth of a gallon. Still, in areas like those Anna Jameson was travelling through, where, as she says, "they came upon us in swarms, in clouds, in myriads, entering our eyes, our noses, our mouths, stinging till the blood flowed," it is possible to sustain as many as 9,000 bites a minute, and at that rate it wouldn't take more than two hours for mosquitoes to account for a greater blood loss than is recommended by the Canadian Red Cross.

But Jameson erred when she referred to her tormentors as "he." In fact, only female mosquitoes suck blood. Like most insects, mosquitoes take in enough food as larvae to last through the pupa stage, but both male and female adults require regular meals of plant nectar or aphid droppings (as ants do) to top up their sugar content. But in many species, the female cannot get enough protein from these sources to allow her to produce eggs. This is where we come in. Mammalian blood is a highly efficient source of protein. So is reptilian blood; mosquitoes evolved during the Age of Reptiles, after all, and many blood-sucking mosquitoes must originally have depended on dinosaurs for their protein; Michael Crichton's conjecture, in *Jurassic Park*, that a mosquito trapped in amber (which is fossilized resin) would contain dinosaur blood is not that far-fetched. Scientists have in fact extracted reptilian DNA from 150-million-year-old mosquitoes, though not enough of it to reconstruct entire dinosaurs.

When dinosaurs became extinct 65 million years ago, most species of mosquitoes continued to feed on reptiles, but others expanded their diet to include blood from the Earth's new dominant animals, the mammals. In northern North America, the last Ice Age wiped out most of the larger mammals, and many mosquitoes turned to humans out of necessity. Our blood contains less isoleucine – the amino acid needed by breeding female mosquitoes – than that of most other animals, but in the north we are much more accessible, partly because for the past five centuries we have been busy reducing the number of alternative mosquito hosts. Certainly in the city, there are more of us than any other kind of blood-protein source.

Our word "mosquito" comes from the Spanish *mosca*, which in turn comes from the Latin *musca*, meaning "fly" (a mosquito is a little fly, Angelo Mosca is a big fly). It was the Spanish who first encountered the little devils in North America, during their attempt to subjugate the peoples of southern and Central America; it was mosquito-borne yellow fever, not the Incas, that drove the Spaniards out of Florida. The hard-bitten conquistadors were right: the mosquito is a true fly, of the order Diptera, family Culicidae. Flies are differentiated from other insects by having only one pair of wings (*di* = two, *ptera* = wings), although they have the vestiges of a second pair, in the form of two small stumps, called halteres, located behind the first pair. Each haltere is equipped with highly evolved sensors that function as flight stabilizers, informing the insect's brain of any diversion from the horizontal, somewhat in the manner of the inner ear in mammals. Apart from the wings, mosquitoes are built like any other insect – tripartite body (head, thorax, abdomen), six legs, two antennae. The thorax is in three sections, and each pair of legs is attached to a different section. The middle section is larger and more muscular than the other two, since it is to this mesothorax that the wings are attached. Mosquitoes in flight move their wings about 600 times a second, which is why they sound more like a whine than a drone: bumblebees move their wings only about 300 times a second.

There are more than 3,500 species of mosquitoes in the world, most of them in the tropics, with an average of 18 new species being discovered every year. Canada has seventy-four species, fifty-seven of which are found in Ontario. Most of Ontario's mosquitoes are blood feeders, but not all blood feeders require human blood. In fact, one of our most common mosquitoes – *Culex pipiens* – gets its blood entirely from birds. There is a species of mosquito that feeds exclusively on the blood of loons (which may explain why loons dive so often), and another that feeds only on turtles, and yet another that cannibalizes the blood-filled abdomens of other mosquitoes. But most mosquitoes in the Toronto area are out after our blood.

Apart from their nuisance factor, mosquitoes are important carriers, or vectors, of disease: as Sy Montgomery writes in *Nature's Everyday Mysteries*, "Mosquitoes kill more humans than any other animal on Earth." More than 100 serious diseases are transmitted by mosquitoes, including malaria, dengue, filariasis (or elephantiasis), yellow fever, and encephalitis. All these diseases with the exception of filariasis are now known in North America. It is no idle claim that mosquitoes have been responsible for major alterations in the course of human history. Mosquito-borne yellow fever halted Spanish exploration in the Americas, prompting later colonists to import blacks to work on mosquito-infested plantations in the Deep South, the theory being that, coming from Africa, they would be more tolerant of mosquito-vectored diseases. In Canada, mosquitoes delayed settlement of parts of British Columbia, such as the Lower Fraser Valley, until controls were instituted in the 1920s. The controls didn't work, but settlers moved in anyway because they thought mosquitoes were now the government's problem.

The two most common mosquito genera found in eastern Canada are *Aedes* and *Anopheles*. The *Aedes* mosquitoes are by far the most numerous, with hundreds of species worldwide and thirty-four in Canada. They originally fed on rodent blood and still do, but have expanded their diet to include all mammals. They prefer to breed in shallow, temporary pools, such as ditches, rain puddles, rain barrels, and discarded car tires. It was *Aedes aegypti* that was responsible for the yellow fever and dengue that did in the conquistadors – fortunately, no *A. aegypti* are found in Canada, but several other *Aedes* mosquitoes, including those with such suggestive names as *A. canadensis, vexans, stimulans,* and *excrucians,* are quite common in southern Ontario and have been known to develop high levels of infection when injected with viruses in the lab.

Few people without electron microscopes take the trouble to distinguish between species of mosquitoes, so I'm not going to dwell too heavily on the minute differences, such as spots on wings or differently arranged scales on legs. But there are some behavioural distinctions that are discernible and interesting, and even important from the point of view of control. Most mosquitoes lay their eggs in water; *A. vexans,* however, lay theirs in the fall in the ground in places where they somehow know there will be water in the spring. When the water comes and reaches a temperature of 8 to 10°C – usually in late May – the eggs hatch within two hours into larva, or "wrigglers": researchers have counted up to 80 million of them in a single hectare of water-soaked pasture. *Aedes* eggs can wait a long time for water. In 1920, Canadian entomologist Eric Hearle dug up a spadeful of earth from the Sumas Prairie in British Columbia, kept it in his lab for a year until it was as dry as powder, then soaked it in water: within five minutes, 653 wrigglers floated to the surface.

After wriggling around in the water like aquatic inchworms for a few days, during which they pass through four distinct stages of growth, feeding on such microscopic organic matter as algae, protozoa, and bacteria, and coming to the surface from time to time to breathe through tubes located near their thin ends, the larvae develop into pupae. The pupae are somewhat larger than the larvae and are less active, floating for the most part just below the surface of the water looking like little inverted commas, or tiny brown shrimp. When disturbed, they too will straighten up and wiggle down out of harm's way, but soon come back to the surface to resume the single-quote position. The adult emerges in a kind of air bubble, like Aphrodite from her conch, so that it doesn't get its wings wet, and then immediately flies off in search of a meal of plant nectar.

It is while supping on nectar that mosquitoes fulfil their second purpose on Earth, that of pollinating wildflowers. Mosquitoes are among the earliest insects to appear in the spring, and so it is their job to pollinate the earliest spring flowers. Some plants have adapted to mosquito pollination by emitting a form of carbon dioxide that attracts the insects to its stamens. One of these is the white trillium (*Trillium grandiflorum*), the floral emblem of Ontario, which ought to endear the mosquito to every resident of Ontario.

After topping up on nectar, the male mosquitoes go into a swarm: thousands of mosquitoes, all males, all hovering in frenzied anticipation of the appearance of a female. The female, attracted by the sound of the males in swarm, flies into their midst, upon which she is immediately beset by a host of males, all of whom can distinguish the sound of her wing beats from those of the thousands of males in the swarm. The female mates with one or more of them by hovering while the male flies beneath her, assuming something like an inverted missionary position, and then she sets off in pursuit of her first blood meal.

A.E.R. Downe, a Canadian entomologist who has been studying mosquitoes for thirty-five years, remembers that when he first became interested in mosquito mating habits, it was assumed that female mosquitoes were monogamous. "Everyone thought that," he says; "monogamy among animals was the catchword of the day. Well, I thought that was ridiculous." Since female *Aedes* mosquitoes lay only one batch of about 546 eggs in their lifetime, says Downe, the idea that an insect as successful as the mosquito would invest its entire future in a single mating session seemed to him to be a trifle anthropomorphic. In nature, putting all one's eggs in a single basket is not good evolutionary strategy. Downe and his colleagues began to look closely at mosquito reproductive habits and, in the process of debunking the monogamy myth, discovered several other interesting things as well. Female mosquitoes, for instance, did not take blood meals until after they had mated, which suggested to them that something in the males' sperm triggered their appetite for blood. Sure enough, they found that recently mated females had high levels of peptide – a protein compound made up of about fifty amino acids – in their guts, which they received from males during mating. When the peptide level was high, the females would take blood meals and lay eggs. In species that laid more than one batch of eggs in a season, the females would not mate again until the peptide count had dropped below a certain level; this drop seemed to trigger a peptide receptor in their heads that told them it was time to mate, feed, and lay eggs again. "We thought that if we could manufacture something in the lab that would block that receptor, then we could prevent the females from mating and laying more eggs, and it would be a rather nice kind of mosquito control. That was our story, anyway, that's how we got money to do the research. But what we were really interested in was knowing more about how mosquitoes regulated their digestive systems."

In northern climates, mosquitoes have had to develop survival strategies for getting through the long, freezing winters. Hibernation – in insects it is called "diapause" – is the obvious answer, but different species of mosquitoes have evolved different kinds of diapause. *Culex* mosquitoes over-winter as adults, sheltering in caves or basements or any dark, quiet place, where they shut down their metabolisms for the long solar night. The buzzing mosquitoes that greet cottagers on their first weekend in May are *C. pipiens*, awakened early from their diapause by heat from the woodstove. Shutting down their metabolism means regulating their digestive systems — which Downe found they do by controlling their production of juvenile hormones — to store food rather than process it. *Mansonia* mosquitoes over-winter as larvae: rather than breathing through tubes at the surface of the water, *Mansonia* larvae bore their respiratory trumpets into the hollow stems of aquatic plants, well below surface freezing level, and remain dormant that way until the pond thaws in the spring. *Aedes* and *Anopheles* mosquitoes over-winter as eggs: the female lays a single batch of eggs in July or August, and these remain unhatched until they have passed through a freeze-thaw cycle, somewhat in the manner of certain flower seeds.

Aedes triseriatus are tree-hole breeders, which means they lay their eggs in hollow stumps, in the crotches of trees, or even in cavities at the bases of trees — anywhere that water will be likely to collect during spring run-off. They will also oviposit (lay eggs) in man-made containers that mimic these natural collectors, such as discarded car tires or tin cans containing water and organic debris, which means they are extremely adaptable to life in the city. *Triseriatus* are not among Toronto's most numerous mosquitoes, but their name often comes up during discussions of mosquito-borne diseases, because in some parts of the United States they are known vectors of the virus that causes California encephalitis. The virus, called La Cross (or LAC), is carried by chipmunks and squirrels and is transmitted to humans by female mosquitoes that take a blood meal from those rodents, become infected themselves with LAC, and then, during a later egg-laying sequence, take a blood meal from a human, transferring the LAC virus to their new host through their anticoagulating saliva. Female *triseriatus* mosquitoes can also transmit LAC and other viruses to their offspring and so allow the disease to over-winter in their eggs to return the following spring, which is how an isolated incidence of a disease can quickly become an epizootic.

This has already happened in Ontario. The encephalitis that hit southwestern Ontario in 1975 was known as St. Louis encephalitis (SLE), an extremely virulent and much more serious form of the disease than California encephalitis. The virus was first identified in 1933 in St. Louis, Missouri, where it caused 266 deaths; forty-three people died from it in St. Petersburg, Florida, in 1962; it showed up in Toronto and Windsor in 1975 and was still around in 1977. *Triseriatus* does not carry SLE – at the time, massive campaigns were launched against C. *pipiens* mosquitoes in the Windsor area, but since C. *pipiens* feed entirely on birds, the campaign served little purpose. Six people died in Windsor and two in Toronto; my wife Merilyn, who was living in Chatham, Ontario, at the time, remembers covering her year-old son's crib with netting at night and being afraid to let him play outside during the day – not only because of the mosquitoes, but also because of the airplanes spraying DDT-laced oil on the lowlands near her house.

South of the border, the SLE vector turned out to be C. *nigripalpus*, an indiscriminate biter that feeds on everything from frogs and turtles to birds and humans, but is not found in Canada. How did SLE get to Canada, then? The theory at the time was that SLE was contained in the blood of migrating birds. Among the birds known to have received the encephalitis virus from *nigripalpus* are snowy egrets, cardinals, mockingbirds, mourning doves, grackles, and blue jays. One of the scenarios offered by immunologists is that a migrating bird, say a mourning dove, was infected with the disease by a C. *nigripalpus* mosquito while wintering in Florida, and then after migrating to Toronto in the spring was bitten by a C. *pipiens* mosquito here.

Downe finds this explanation "highly doubtful," because C. *pipiens* has never been known to take blood meals from humans, although he admits that anything is possible with mosquitoes. His own guess is that the vector was C. *tarsalis*, a very common mosquito in Manitoba, Saskatchewan, and Alberta, and a known vector of Western encephalitis (WE), a mild form of the disease that affects mostly horses and chickens, but that can be passed on to humans. In studies in Alberta, he says, four people out of ten had had WE without knowing it. Like all *Culex* mosquitoes, *tarsalis* over-winters as an unmated adult and is thus one of the earliest mosquitoes to emerge in the spring. For a long time it was not known how the encephalitis virus survived the winter; it couldn't have been in *tarsalis* females, because they don't take their first blood meal until spring, and they do not inherit WE from their parents.

In 1964, Downe and a colleague from the University of Alberta studied the relationship between C. *tarsalis* and red-sided garter snakes, *Thamnophis sirtalis rudix*. The snakes over-wintered in the same caves and mine shafts as the mosquitoes, and Downe wondered if the encephalitis virus wasn't lying dormant in the snakes. "We went into a few abandoned mine shafts and hauled out garbage pails full of garter snakes," says Downe, "brought them back to the lab, fed them with tweezers on little bits of tinned smelt, took blood samples from them, and tested them for WE. All that summer we got nothing, zero, and we'd just about given up on them. Then in the fall somebody who thought we were going to throw the snakes out put them in the refrigerator overnight, just to get them out of the way. The next day we gave them one last blood test and bingo! the snakes were hot with virus." Downe thinks it took a shock to their system, like the sudden cold temperature in the refrigerator, to release a shot of adrenaline that proliferated the virus, and figures that when the snakes emerge from their hibernacula in the spring they are ripe with encephalitis. The mosquitoes, emerging from diapause about the same time, bite the snakes for their first blood meal, take up the disease, and then pass it on to other hosts throughout the summer.

C. *tarsalis* mosquitoes are not common in eastern Canada, but specimens have been found in Windsor, Kitchener, and Toronto. They have also been experimentally infected with LAC virus, which they have retained, and Downe says he'd bet a week's salary that *tarsalis* will also prove receptive to the St. Louis strain. "Of course," he adds, "now that I'm retired, a week's salary doesn't amount to much, but all the same, I'm pretty sure that *tarsalis* is the culprit."

There is a new candidate for the spread of disease in Canada in the near future, an *Aedes* mosquito from Asia that has only recently turned up in North America. *A. albopictus*, also known as the tiger mosquito because of its beautiful black-and-white coloration, was first discovered on this continent in 1985, in Houston, Texas. Investigators determined that it was being imported in used car tires from Japan by North American retread companies. In the past decade, *albopictus* has spread north and west, and is now found in nearly twenty states; in some cities, it is already the most common biting mosquito. *Albopictus* is a carrier of dengue fever in Asia and is now known to carry the LAC virus in the United States. In 1988, researchers with the U.S. Centers for Disease Control tested 6,159 *albopictus* mosquitoes from Illinois, Indiana, and Ohio and found no LAC virus; one year later, they tested another 14,000 *albopictus* from the same states and found they carried seven different virus strains, including LAC.

This sudden ability to harbour viruses unknown to them in their native habitat is a striking example of mosquitoes' ability to adapt to new environments; any organism that produces 500 offspring a year is bound to produce one or two that have characteristics favourable to survival in a new situation. Dr.

Downe points to an even more striking example: when *albopictus* first arrived in Texas, its eggs were timed to go into diapause at a date appropriate to that southern latitude, which was more or less the same as its Asian origins. Farther north, it was unable to over-winter, which kept the species from spreading into Canada. "Now," says Downe, "horror of horrors, the specimens they're collecting in Chicago have evolved the ability to over-winter as eggs and can therefore survive our cold temperatures." So far, says Downe, no *albopictus* mosquitoes have been reported in Canada – their northern march has halted in southern Michigan – but he feels that their arrival in southern Ontario is just a matter of time. When they do show up here, *albopictus* could become the most common *Aedes* mosquito in Toronto, and immunologists could have a whole new encephalitis vector to deal with.

This seems an appropriate place to point out that mosquitoes do not transmit the AIDS virus. So-called "pool-feeders," like black flies, horse flies, and deer flies, which actually bite a chunk out of their victims and then dunk their heads and forelegs into the blood that wells up from the puncture, can pass diseases on from host to host, in much the same way that dirty hypodermic needles can transmit AIDS. But because mosquitoes penetrate their host's skin cleanly and transmit whatever virus they carry through their salivary glands, they actually have to contract a disease themselves before they can pass it on to another organism. And so far no mosquito has tested HIV-positive: the virus quickly dies in the mosquito's gut without replicating, which, come to think of it, ought to suggest an interesting line of enquiry to AIDS researchers.

During the construction of the Rideau Canal, which began in 1828, hundreds of workers died from a disease they called "swamp fever." The problem increased whenever a section of the canal was routed through low, swampy land and was relieved when they stayed up on higher, dry ground. There

was something in the water – some kind of gas, they thought
– that got into the workers' lungs and brought on high fever,
fits of trembling, sweating, and eventually death. The disease
was correctly identified as malaria – the name comes, after all,
from the Italian *mal aria*, meaning "bad air" – but at the time
nothing was known about how the disease was transmitted.
Unfortunately, very little of the Rideau Canal passed through
high ground, and so all along the system you can still find
small graveyards with white, indecipherable headstones
marking the graves of the 500 anonymous men – one-quarter
of the entire work force – who succumbed to Upper Canada's
bad air.

It wasn't until much later that scientific historians realized
that Colonel By's doctors were partly right in their assessment: the disease that killed the workers did indeed come
from the swamps through which they worked; but it wasn't
swamp gas that carried the disease, it was mosquitoes, probably of the species *Anopheles quadrimaculatus*. In the tropics,
Anopheles mosquitoes are the principal vectors of the single-
celled protozoans that cause malaria, but we don't normally
think of Canada as a locale for malarial epidemics; the disease
has a Raj ring to it, associated with white-mustachioed British
officers sipping gins-and-tonic (for the quinine, don'cher
know) on netted verandahs and surrendering periodically to
fits of uncontrolled shuddering. This is a modern conceit: it's
true that the more than 100 million new cases of malaria that
are reported each year are in the tropics, as are the more than
one million people a year who die from it, but the days when
North American children were force-fed shot-glasses of War-
burg's Tincture of Quinine every morning have only recently
passed: John Kieran, author of *The Natural History of New York
City*, recalls those tinctures with a shudder that mimics the
disease they were supposed to prevent: "Despite this daily
dosage," he writes, "we all suffered considerably from recur-
rent malaria, and so did most of our neighbours."

A. quadrimaculatus is still with us and is in fact one of Toronto's most common mosquito species. The best way to distinguish it from the *Aedes* mosquitoes is to observe it during its larval stage. First off, *Anopheles* females prefer to deposit their eggs in permanent bodies of water, such as swamps, ponds, and river margins, rather than in temporary puddles, or mud, or even in the bark of trees, as *Aedes* tend to do, although *Anopheles* larvae have been found in polluted water and in tin cans. Also, when *Aedes* (and *Culex*) larvae aren't wriggling around looking for food, they float in the water in a vertical position, their hairy food-gathering fans pointing straight down, and their "tails" sticking up and curving slightly just at the surface, so that their breathing tubes, or "prothoraxic respiratory trumpets," located one or two segments from their tips, can poke up through the water's surface. The respitory trumpets of *Anopheles* larvae are attached at their thick ends, and they float at the surface in a nearly horizontal position, almost like a straightened-out pupa, with their trumpets sticking up from what looks like their shoulders. Adult *A. quadrimaculatus* mosquitoes can also be distinguished from other species, if you have a large enough magnifying glass, by the presence of four black dots on each wing. They are night feeders, but have been known to take blood meals in cool, darkened buildings, such as barns – they prefer cattle blood, but will take human blood as a poor second – and over-winter as adults in basements and other such places.

We have gained our victory over malaria and other mosquito-borne diseases at an incredible cost. Even the most benign methods of mosquito control – draining swamps and lagoons or coating them with oil – have had immeasurable negative effects on other species of plants and wildlife that depend on those wetlands for their survival. And draining marshes didn't work all that well against mosquitoes, whose eggs can lie dormant in dried mud for up to five years. But more aggressive methods of mosquito control have proven to be much worse.

New insecticides have come on the market – Agriculture
Canada lists thirty-three active compounds registered for use
against mosquitoes. Larvicides such as temephos and adulti-
cides such as malathion have proven effective, but they, too,
have drawbacks similar to DDT: they are harmful to non-tar-
get species, and target species have developed resistance to
them. A study conducted from 1985 to 1989 in Maryland
showed that all mosquitoes there were still susceptible to
malathion, but that *C. pipiens* was resistant to temephos. Similarly, the newest pesticide to come down the pipe – *Bacillus
theringiensis israeliensis*, or Bti, a bacteria that enters the mosquito's gut and causes it to stop eating (and therefore die) –
has been in use only for the past decade, and already *C.
tarsalus* – the principal WE vector – is showing resistance to it.
And there is some evidence that Bti is harmful to other
aquatic species, such as wood frogs.

"Whenever anyone asks me about insecticides," says
A.E.R. Downe, "I always urge caution. We just don't know
enough." Downe tells people to protect themselves individually rather than try to eradicate an entire species. "Don't wear
dark blue, I tell them," he says. "Especially blue jeans: for
some reason, and I don't know what it is, mosquitoes love
blue-jean blue. It took me a long time and a lot of scratching
before I figured it out, but now I wear only yellow or orange
coveralls during field studies. They don't actually see colour,
but they seem to react to something in colour."

The oldest defence against mosquitoes, used by native
peoples and early settlers, was the smudge fire, "little fires to
windward of the house," wrote P.H. Gosse in 1840, "covered
with wet chips and earth, which, smothering the flame, make
a dense smoke; this being wafted by the wind around the
house, prevents the approach of the flies, as they cannot
abide smoke; so we tolerate one inconvenience to dispel a
greater." Out West, where mosquitoes in lumber camps in the
1910s were bad enough to kill mules, work was either halted

for weeks on end or else conducted in the continual murk of netting and smoke. Here is Eric Hearle's account of the conditions he encountered in British Columbia:

> In 1911, an electrical plant for generating 40,000 horsepower was under construction at Stave Falls, near Ruskin. Mosquitoes appeared in intolerable numbers and it was feared that it would be impossible to retain the men at work.... The officers of the company persuaded the men to remain by guaranteeing that everything possible would be done to alleviate conditions. Over $2000 were expended in mosquito netting, on smudges and other methods of protection. Every man was provided with a mosquito veil and gauntlets, but even with such protection the efficiency of the labourers was greatly reduced. The men were obliged to keep their veils on even at meal times, and smudges were kept burning in the bunk houses night and day.

Settlers here in the east discovered the virtues of the wild-flower pyrethrum, which repelled mosquitoes and black flies so effectively that they grew it in their gardens and hung bunches of it in their doorways – today, the compound pyrethrin is the chief ingredient in those mosquito coils people burn in their cottages and on their patios. Sue Hubbell, in her wonderful book on insects, *Broadsides from the Other Orders: A Book of Bugs*, notes that Avon's Skin-So-Soft is the "universal mosquito dope for country people," but University of Guelph professor Doug Surgeonor, who tests mosquito repellents for the Canadian military, says that remedies such as skin creams, bug zappers, and sonic booms might work for short periods, but long-term commercial deterrents like N-diethyl-m-toluamide, otherwise known (for some reason) as Deet and the only ingredient in the over-the-counter, Canadian-made repellent Muskol, will keep mosquitoes at bay for up to eight hours. The label on my bottle of Muskol warns that its contents "may damage furniture finishes, plastic, painted surfaces and synthetic fabrics," and a few years ago the Environmental Protection Agency was taking a close look at reports that Deet caused eye irritation, dermatitis, and even brain damage in infants. The EPA eventually cleared Muskol for sale in the United States, but Merilyn – who makes her own mosquito repellent from citronella mixed with baby oil – says, "Citronella lasts for only an hour or so, but I'd rather apply citronella to my skin every hour than douse it once a day with something that will melt plastic." *Mother's Remedies*, a delightful compendium of "over one thousand tried and tested remedies from mothers of the United States and Canada," published in 1910, recommends that post-exposure mosquito victims "remove the sting in the wound" (mosquitoes don't leave a sting in the wound) and adds that "diluted vinegar applied to the bites is sometimes of help. Camphor is also good."

Everyone, it seems, has an attitude towards mosquitoes. Perhaps, in the end, Gosse has the most sensible attitude of all, at least from an environmental point of view: "There is no other help," he concludes, "but patience."

OF THINGS RANK
AND GROSS IN NATURE

One day in early June, I spent an afternoon on my hands and knees in our back yard, digging weeds out of the lawn. It is a small lawn. An earlier owner covered a third of the yard with a concrete slab that at one time supported a garage, and most of the rest has been converted by us to vegetable and perennial beds. But there is enough grass left to merit the kind of inattention I was giving it, and before long I had a bushel basket full of dandelions, plantain, creeping Charlie, and white clover, and a scruffy patch of grass that would stir the heart of the most ardent monoculturist.

Our neighbour's daughter wandered over and, in the way six-year-olds have of getting directly to the point, asked me what I thought I was doing.

"Digging weeds," I said.

"What's a weed?" she asked.

"A weed is a wild plant that grows where we don't want it to grow," I said.

"Why?" she asked.

"Why what?" I replied.

"Why don't we want it to grow there?"

"Because we want something else to grow there."

"Like grass?" she said.

"Yes, like grass," I replied.

"Why?" she asked.

Lisa pretty much summed up my own thinking on the subject. My own thinking almost always begins with a statement and ends with a question, or a whole raft of questions, instead of the other way around. After a week of reading and looking, I'm still not sure I know what a weed is, or why we prefer grass to, say, dandelions. But I find some comfort in knowing now that nobody else does either.

"What is a weed?" asked J. Eaton Howitt, lecturer in botany at the Ontario Agricultural College in Guelph and author of *Weeds of Ontario*, the Ontario Department of Agriculture's 1911 handbook for farmers. He then went on to give three definitions that added up to the same fuzzy answer I gave Lisa: "A plant out of place," he said, or "Any injurious, troublesome or unsightly plant that is at the same time useless, or comparatively so." Or "A plant which interferes with the growth of the crop to which the field is temporarily devoted."

Well, does this mean that if I took a dandelion seed and deliberately planted it in my lawn because I liked the look of it, or in my garden because I wanted to eat it, it would no longer be a weed? Clearly, the definition of "weed" is a matter of personal preference and has very little to do with the behaviour of plants in nature. What is and is not a weed has pretty much been determined by farmers and gardeners, and asking them to define a weed is like asking a chicken farmer to define a coyote. Keeping certain plants out of an area intended for certain other plants may be necessary in the country, where their presence may interfere with a farmer's ability to feed a family, or a hundred families, but it may also be that in the city, where plants have a utility beyond commerce, we can afford to be more tolerant. After all, some people may think that any plant growing in a city is "out of place," and some cities have bylaws forbidding citizens to grow anything but grass and domestic flowers on private property. But there are others who think that an abandoned yard full of wild flax and buttercups is a lot more attractive than an abandoned yard full of broken bricks and pieces of aluminum siding, and I'm one of them. It is in the nature of wild plants to take over waste places, and cities provide a lot of waste places.

Take the Leslie Street Spit, for example. Begun in 1959 as a project that had some vaguely hopeful connection to the St. Lawrence Seaway (more ships would be coming to Toronto, so we needed more docking space), the Spit is really nothing more ambitious than a huge landfill site protruding five kilometres out into Lake Ontario. But by 1980, it had been colonized by 278 species of plants, from wild grasses and sedges to full-grown trees and shrubs. How did all those plants get there? Many of them were trucked in with the contruction waste, as bulldozers cleared former residential districts to make room for businesses, but most of the plant seeds must have travelled the way seeds are supposed to travel: borne on the wind, washed ashore by the waves, deposited in bird droppings – some 200 species of birds have also been seen on the Spit at one time or another – or stuck to the fur of animals or the socks, pant cuffs, or boot soles of us human vectors. Some of the plant species are native to the Toronto area (about 122 of them), but most are not, having been introduced either intentionally or accidentally from some other part of North America, or from some other continent. Dandelion, Canada thistle, plantain, mullein, white and red clover, lamb's quarters, pineappleweed, coltsfoot, Queen Anne's lace, shepherd's purse, common mallow, creeping Charlie – all are European or Eurasian plants brought over by settlers either as garden vegetables (lamb's quarters, for example, a near relative of spinach, or yellow rocket, also known as winter cress), or accidentally mixed in with grain seeds (dodder seeds are virtually indistinguishable from those of alfalfa), or cattle feed (wild mustard – *Sinapis arvensis*), or just stuck in the tread of an old boot. One of the more interesting plant histories associated with the Spit is that of the alkali-grass, *Puccinellia distans*, a salt-favouring plant not found anywhere else in Ontario – its natural home is the salt marshes of New Brunswick – but which flourishes at the base of the Spit, just to the left of the entrance gate, where city snow-removal trucks stockpile snow during the winter. The ground there is so saline that not much else will grow in it. The speculation is that tourists brought seeds of this grass stuck in the tires of their camper vans after driving through the Maritimes.

Leslie Street Spit is a great place to go to get an idea of what would happen to Toronto if the city were allowed to run wild. It wouldn't look anything like it looked before we got here. Originally, the section of the north shore of Lake Ontario between the Rouge and Humber valleys that is now Toronto was part of a life zone called the Carolinian forest, a dense bioregion overgrown with cedar, pine, and mixed hardwood forest, the ground covered with ferns and marshland plants. We have since changed it to something resembling a subtropical savannah, as our species is wont to do. This is a huge and significant change: eight out of every ten species of plant that grows here now has been introduced from somewhere else.

Most of the grass in Toronto, for example, has been brought in in the form of lawn, or turf grass. The three commonest species are Kentucky bluegrass, perennial ryegrass, and red fescue: these three grasses, mixed together in a lawn, guarantee early greening, hardy wintering, and resistance to being walked on. We North Americans like our lawns. The Kentucky-based American Turf Institute recently estimated that if all the lawns in the United States were gathered together into one huge lawn, it would cover an area equal to all of New England. The institute also noted that four times the amount of pesticide is used on lawns as on agricultural crops in North America.

This has not been entirely beneficial, either to native species or to the environment as a whole. Lawns consisting entirely of imported grass species (Kentucky bluegrass, despite its name, is a European variety) require more water and more frequent cutting than any other kind of groundcover – our native buffalo grass, for example, looks like turf grass but never grows higher than fifteen centimetres and never requires watering. And the average adult in North America spends thirty hours a year pulling up dandelions, riding gas-guzzling power-mowers, and setting up lawn sprinklers to keep his or her lawn looking as spiffy as the eighteenth green at Glen Abbey.

What is grass, anyway? There is, of course, far more to it than that chrome-green stuff we see on our lawns. It may be that the biblical observation that "all flesh is grass" (Isaiah 40:6) is meant to imply that life is as transient as a blade of annual Kentucky bluegrass, and that we are all destined to be cut down by Mad Meg riding a Lawnboy and tossed into the Great Compost Bin in the Sky, but it is possible to look at it more literally: grass is the food that feeds all flesh. We are what we eat, and we all eat grass, either directly (as grain) or indirectly (as grain-fed meat). For instance, the rank in importance of sugar cane, wheat, corn, and rice (all grasses) in the world's major food crops is 1, 2, 3, and 4. Two more grasses, barley and sorghum, are 7 and 11. Of the 3.5 million tonnes of food provided by the top eleven crops, grass accounts for 2.5 million tonnes, or 70 percent. And all of the red meat and much of the white meat we consume comes from animals that are natural grass- and seed-eaters. Take away grass from the Earth, and we all starve. We might also be naked and illiterate: papyrus and linen are both grasses. "Surely," as Isaiah goes on to say, "the people is grass."

There are, worldwide, 4,500 species of grasses, all of them belonging to the Gramineae family of herbaceous plants and characterized by having cylindrical, hollow stems furnished with knots or nodes. At each node, a leaf that begins as a sheath wrapped around the stem splits and protrudes from the stem in the opposite direction to the leaves immediately above and below it. The leaves themselves are long, slender, and pointed (which is why they're called "blades"), and the flowers are arranged in spikes (like timothy), clusters (like oats), or panicles (as on lawn grass). Look at your neighbour's overgrown lawn in early June: that faint, white spray waving just over the green is a million tiny grass flowers. Each flower is made up of tinier spikelets, which begin at the base in two independent and sterile bracts that form around the ovary, stigma, pollen, and stamen. This basic structure is common to all grasses, including such apparently diverse forms as corn, sugar cane, sweetgrass, bamboo, and rice, as well as to the more familiar fescues, bluegrasses, ryes, and bent grasses found on most lawns.

What does grass do? Of the world's 4,500 species of grasses, only a handful are agricultural or ornamental: the rest just grow wild. Wild grasses play an important role in an ecosystem. Left to their own devices, grasses grow where trees and shrubs generally do not. Large plants require huge amounts of water, which is why they send their roots deep down into the aquifers. A full-grown maple tree will transpire 227 litres of water through its leaves into the air every day, and flush another 227 litres into the ground through its root system every night – incidentally watering whatever grows beneath its spreading boughs. In dry areas, where the ground-water table is too low and rainfall is too infrequent (i.e., seasonal) for large trees, grasses move in and protect the soil. Without grass, the sun would dry up the earth and the wind would blow it away. Or else seasonal rains and spring floods would wash it away. Grass holds some water in the top several centimetres of soil (for its own use), and directs the rest of it (about 70 percent) into streams or lakes, where it collects and might, en masse, do some good to more deeply rooted plants.

In arid regions this is a decidedly beneficial function. Grasses in arid regions make trees possible elsewhere. Grasses in arid regions also photosynthesize sunlight, thereby reducing atmospheric carbon dioxide and producing oxygen in parts of the Earth that otherwise would not do this. In fertile areas, however, where the natural tree cover has been removed, grasses move in (or are brought in) and continue to perform the functions their DNA tells them to: they prevent rainwater from penetrating down into the aquifers by directing most of it elsewhere, into streams or lakes or, since we are talking about urban areas, storm sewers. This is decidedly not a beneficial function. The American architectural iconoclast Malcolm Wells, in his book *Gentle Architecture* – in which he advocates, among other things, building houses, office complexes, and even whole communities underground – disdains grass, by which he means lawns, as just another form of pavement. As more and more land is turned into suburban building sites, he says, it "becomes very efficiently paved, if not with blacktop or concrete or roofing materials, then paved with closely mowed turf – lawn grass – which is no slouch as a paving material either. Neatly trimmed grass can be counted on to repel almost half as much rainwater as a shingled roof."

A lawn might be defined as grass growing where it wasn't meant to. The word, originally spelled "laund," is from the French *lande*, meaning moor, and was first applied in England not to cultivated patches of turf but to naturally open spaces in forests, i.e., glades or meadows, grassy swards, open moors, and other places where trees didn't grow. Having a "lawn" on one's property meant that one owned enough woodland to encompass a meadow and was wealthy enough to own *lande* without having to convert it to agricultural *land*. As cities grew in importance and size, they began to name their parts for what they replaced (we still do this, with suburbs called

Greenwood, streets called Pinecrest and Orchard View, and shopping malls named Meadowvale), and little squares of tough, trodden, desperately tended fescue in front of working-class row-houses were dignified by being called "lawns."

Our own fondness for lawns can be traced back to Lancelot "Capability" Brown, the eighteenth-century British landscape architect who quite literally created what we now think of as the English countryside. Brown's mission was to disabuse British estates of the Gothic temples, Delphic grottoes, and Palladian bridges that had been imposed on them by classically minded architects of the past, and to replace them by what he called "pure" landscape: long, rolling, verdant vistas, grassy hills dotted here and there with clumps of trees for perspective, and in the distance a ridge of forest, as a framework for the whole. Turf grass of course existed before that, but not on such a grand scale, having been confined to tennis courts (in Elizabethan England), or growing on small hummocks meant to be used as benches, or to footpaths between beds of more exotic imports in domestic gardens. Capability Brown used grass as a painter uses green paint, to fill in the blank spaces between centres of focus. He bestowed his meadowed vision on 140 estates in his lifetime, and when we think of rural England today – in fact, when we think of beauty in landscape today – we are more or less thinking what Capability Brown told us to think. In his book *Biophilia*, Edward O. Wilson quotes Captain R. B. Marcy, leader of a U.S. government expedition to the midwestern prairies in 1849, who declared the landscape at the headwaters of one of the forks of the Brazos River to be "as beautiful a country for eight miles as I ever beheld." What he beheld was a natural replica of a Capability Brown landscape:

> It was a perfectly level grassy glade, and covered with a growth of large mesquite trees at uniform distances, standing with great regularity, and presenting more the

appearance of an immense peach orchard than a wilderness. The grass is of the short buffalo variety and as uniform and even as new mown meadow. . . .

Marcy found the scene beautiful because it conformed to
what his eye had been conditioned to identify as beautiful: a
"perfectly level grassy glade," clumps of trees "at uniform distances," a view that looked more like "an immense peach
orchard than a wilderness." This is the English countryside as
created by Brown. Similarly, Alexander Ross describes his
first view of the Red River Settlement in 1856 "one of the
finest views and most fascinating prospects in nature." What
did he see there? A Capability Brown construct: "On the east
side the landscape is more varied, with hill and dale, and
skirted at no great distance by what is called the pine hills,
covered with timber, and running parallel to the river all the
way." This is the concept of natural beauty the British
colonists brought with them along with their axes and their
ploughs.

Those who came to Ontario, however, and particularly to
the Toronto area, didn't find much in the way of English
countryside. What they found were dense, constricting pine
forests, pestilential swamps, unfamiliar and disconcerting
mixed bush. Anna Brownell Jameson described a journey
through the woods north of Toronto (i.e., somewhere near
today's Yonge and Eglinton) that "strongly impressed and
excited" her:

The seemingly interminable line of trees before you; the
boundless wilderness around; the mysterious depths amid
the multitudinous foliage, where foot of man hath never
penetrated, – and which partial gleams of the noontime
sun, now seen, now lost, lit up with a changeful, magical
beauty – the wondrous splendour and novelty of the
flowers, – the silence, unbroken but by the low cry of a

bird, or hum of insect, or the splash and croak of some huge bull-frog, – the solitude in which we proceeded mile after mile, no human being, no human dwelling within sight, – are all either exciting to the fancy, or oppressive to the spirits, according to the mood one may be in. Their effect on myself I can hardly describe in words.

Few early settlers shared even Jameson's hesitant appreciation of this unfamiliar landscape. Samuel Thompson, a journalist, poet, and naturalist who came to Ontario in the 1830s, wrote that the site of the City of Toronto "was nothing but a rough, swampy thicket of cedars and pines mixed with hardwoods," an accurate but unflattering description of a Carolinian forest.

Settlers of Thompson's persuasion quickly set about changing the landscape into something they could understand. They cut or burned down the trees, cleared the land for agriculture – British-style agriculture, which meant large fields of cleared land, with here and there a clump of trees or a stone pile, and in the distance a ridge of forest for firewood and maple syrup. Take a look at any Ontario farm today, and you will see a Capability Brown landscape. Around their houses they recreated the gardens they had left behind in Europe, importing seeds of familiar vegetables and flowers, conditioning this foreign soil to accepting them, and thereby changing the landscape forever. Hardly a single letter from an Ontario homesteader to relatives in Great Britain does not thank someone for some new packet of seeds. "We have got quite a nice garden this year," Anna Leveridge wrote to her mother in the 1880s from her homestead in Hastings County:

and if the crops do well, shall have quite a harvest. The high hill in front is coming green with wheat, and in the

valley below barley is flourishing, while both hill and valley is laid down in seeds for a pasture for another year.
Then comes a broad strip of peas for the pig, then a large
sowing of garden peas for ourselves. . . . Then I have a
large piece of beans for use both green and dried, as well
as several long rows of broad beans. Then comes a large
bed of onions. . . . Then I have beds of parsnips, carrots,
radishes, and lettuces, and my strawberries are in blos-
som. Behind the shanty is a little piece of clover sown last
year which keeps our pig. I have some flower seeds coming up in front, and am going to plant out this week a
large bed of tomato plants, and cabbages, etc.

All that growth, and not a native plant to be seen. Catherine Parr Traill mentions native grasses and other groundcov-
ers, such as wild strawberries, which settlers found growing
around their shanties, but she does not consider them as anything but ornamental. "One or two species of grass that I have
gathered bear a close but of course minute resemblance to the
Indian corn," she wrote to her mother in 1832, "having a top
feather and an eight-sided spike of little grains disposed at the
side-joints. The *Sisyrinchium*, or blue-eyed grass, is a pretty little flower of an azure blue, with a golden spot at the base of
each petal; the leaves are flat, stiff, and flag-like; this pretty
flower grows in tufts on light sandy soils." But it was the
wrong kind of grass. It may have been pretty, but Traill's
brother's cattle wouldn't touch it. (Although still called blue-
eyed grass, *Sisyrinchium* is actually a member of the lily family.) Her brother, Samuel Strickland, in his guide for new
settlers called *Twenty-Seven Years in Canada West*, published in
1853, tells prospective newcomers what to expect, and what
to bring with them:

In this country, hay-cutting commences about the first or
second week in July. Timothy-grass and clover mixed – or

timothy alone – are the best for hay, and the most produc-
tive. The quantity of seed required for new land is six
quarts of grass-seed and two pounds of clover to the acre;
on old cleared farms nearly double this seed is required. . . .
We have other sorts of grasses, such as red-top, blue-joint,
&c.: these grasses, however, are inferior, and therefore
never grown from choice.

All the grasses recommended by Strickland were European imports. Creating ideal habitat for them also created
ideal habitat for native invasive plants. Wild plants need disturbed ground over which to spread, and the early settlers in
Ontario disturbed a lot of ground. Their favourite method of
clearing land was to burn down the trees, on the theory that
ash improved the soil. In England, ash improved certain types
of soil by increasing its acidity, but it is by no means a universal cure-all. The untilled soil of Ontario was already highly
acidic; burning made it even more so. And lots of native wild
plants preferred acidic soil. Before the settlers could plant
their own imported crop grasses, those pesky native wild
plants moved in. "The fire-weed," wrote Traill, "a species of
tall thistle of rank and unpleasant scent, is the first plant that
appears when the ground has been freed from timbers by
fire." I like that: "freed from timbers," as if the torch-bearing
settlers were liberating a grateful earth from its burden of
trees. "The next plant you notice is the sumach, with its
downy stalks, and head of deep crimson velvety flowers,
forming an upright obtuse bunch at the extremity of the
branches. . . . This shrub, though really very ornamental, is
regarded as a great pest in old clearings. The raspberry and
wild gooseberry are next seen, and thousands of strawberry
plants of different varieties carpet the ground, and mingle
with the grasses of the pastures."

Did it not occur to anyone that the earth was trying to tell
these people something? Sumach is richer in malic acid than

apples are and has more Vitamin C; native North Americans travelled far for raspberries and gooseberries; we have made the sweetest jam imaginable from wild strawberries. Were these gifts prized by our English forbears? "I have been obliged this spring," Traill continues, "to root out with remorseless hand hundreds of sarsaparilla plants, and also the celebrated ginseng, which grows abundantly in our woods: it used formerly to be an article of export to China from the States, the root being held in high estimation by the Chinese."

None of these plants were much valued by settlers – the trade in ginseng (*Panax quinquefolium*), once vigorous and highly profitable (Frère Marie-Victorin, in his monumental study of the flora of Quebec, *Flore Laurentienne*, compares the discovery of natural ginseng in Canada to the discovery of gold in California: "Settlers found it more profitable to gather ginseng than to plant wheat," he wrote) died out in the eighteenth century because impatient settlers dried the roots quickly in great ovens rather than slowly in barns, thereby reducing their quality. Sarsaparilla (*Aralia nudicaulis*, also called Indian-root) was used by the Indians to make a tisane said to purify the blood, and roots of fireweed (*Epilobium angustifolium*) were steeped to make a liniment against scrofula. Early beekeepers may have savoured fireweed honey, but the plant itself didn't survive invasion by the tropical staghorn sumach (*Rhus typhina*), which, at its flowering in mid-June, announced the true beginning of bee season. Berry-bearing plants weren't actually spurned, but they never became the focus of much interest, and certainly were not cultivated. "Our wild red raspberry," write James Soper and Margaret Heimburger in *Shrubs of Ontario*, "is closely related to the cultivated red raspberry of the garden, a species of European origin." In fact, both have the same Latin name, *Rubus idaeus*. Why did settlers import domesticated *R. idaeus* from Europe when there were perfectly good *R. idaeus* growing wild? Because our English settlers had already decided that anything that grew wild, no matter how useful or beautiful, was by definition a weed.

"A plant is a weed largely by man's definition," writes Anthony Huxley in *Plant and Planet*, "and is often created by his activities." Quite so. If we hadn't imported European cattle instead of taming the bison we found here, we wouldn't have had to import oats and outlaw the native variety. If we hadn't imported oats, we wouldn't have tried to eradicate wild mustard. If we didn't have such an infatuation with the perfect lawn, we wouldn't look upon dandelions as scourges from hell. We might look upon them as the tasty, comely, and highly interesting plants they are. If we weren't.so intent on spinach, a domesticated member of the goosefoot family (Chenopodiaceae), we might not pull up the lamb's quarters (*Chenopodium album*), a wild member of the same family that grows in our garden and is every bit as tasty and nutritious as anything sold in a seed catalogue.

On early mornings in March, lamb's quarters is regularly (if surreptitiously) harvested in the Rouge Valley and sold in Chinese grocery stores on Dundas Street. In our own garden, we pluck it out of the tomato patch but let it grow in with the salad greens (we sow all our salad greens in one blended patch called a mesclun). We don't call it a weed, we call it a "volunteer": other volunteers in our garden include dill, which we haven't planted for several years because it keeps coming up on its own, and purslane (*Portulaca oleracea*), a plant that grows wild – the English writer William Cobbett described purslane in 1819 as "a mischievous weed that Frenchmen and pigs eat when they can get nothing else" – but the seeds of which are sold anyway by seed houses as Portulaca. This must be a highly profitable business, as purslane is one of the most fecund plants known – a single capsule can produce a million seeds – and I have a wry vision of gardeners sowing a bed of Portulaca seeds and weeding out the volunteer purslane that comes up around it.

Designating a volunteer plant a "weed" is one thing, but we don't let it rest there. We single out certain among them for special opprobrium and call them "noxious weeds." In Ontario, this practice goes back to 1887, when the Department of Agriculture issued its first pamphlet – a warning to farmers that perennial sow thistle had spread like the bubonic plague into rural Ontario. By 1892, the list of plants to eradicate had expanded to include Canada thistle, wild flax, pigeonweed, ragweed, couch grass, ox-eye daisy, burdock, blueweed, wild mustard, and wild oats. In 1911, J. Eaton Howitt ended his handbook on weeds with a copy of the Province of Ontario's "Act to Prevent the Spread of Noxious Weeds." Its biblical-sounding injunction is unequivocal:

> It shall be the duty of every occupant of land, or, if the land be unoccupied, it shall be the duty of the owner, to cut down and destroy all Canada thistles, ox-eye daisy, wild oats, ragweed and burdock growing on his land, and all other noxious weeds growing on his land to which this Act may be extended by by-law of the municipality, so often each and every year as is sufficient to prevent the ripening of their seed. . . .

This is an odd list, I think. What's so noxious about ox-eye daisies? "The Oxeye Daisy," Howitt wrote, "is a weed naturalized from Europe, and is very closely related to the Chrysanthemum or national flower of Japan." So? The ox-eye daisy is one of the most beautiful and ubiquitous roadside flowers in North America – it pops up every year through the cracks in the concrete pad in our back yard – in fact, it may have created the cracks – and its radiant white flower has brightened more country roadsides than all the geraniums and petunias ever planted. In June, the median between the lanes of the 401 leading into Toronto is a riot of ox-eye daisy. It is the flower every child has used to find out if he or she is loved, being the perfect flower for it because the number of its petals is not fixed, but varies seemingly at random, making it impossible to cheat by always starting with either "She loves me" or "She loves me not," I could never remember which. But in 1911, the Department of Agriculture declared it to be "most troublesome in pastures," and so anyone found with it on their property was "liable to a fine of not less than $10 nor more than $20."

Ox-eye daisies are now off the official list of noxious weeds, I'm happy to report, and some of the new ones on it make a certain sense: poison ivy (*Rhus toxicodendron*), for example, is a harmful plant that should be avoided (although eradication seems a trifle extreme for a plant whose toxic effects can be treated with a little bicarbonate of soda). The toxic ingredient in poison ivy, by the way, is also found in strawberries and kiwis, so if you are extremely sensitive to poison ivy, you should avoid those fruits. Poison hemlock (*Conium maculatum*) is on the list and should also be given a wide berth: though rare, it is the same plant from which the Athenians prepared Socrates' last drink. But if having poison leaves makes a plant a noxious weed worthy of eradication, then we must also eradicate every tomato, potato, and rhubarb plant in Toronto.

There are some questionable plants on the current noxious-weed list. Wild carrot (*Daucus carota*), for instance, better known as Queen Anne's lace, is on it. A biennial member of the parsley family, it has an edible, carrot-like taproot and a rosette of finely dissected leaves, "as pretty as anything you might plant," says Michael Pollan in *Second Nature*. It is, in fact, the plant from which our cultivated carrot was first derived, probably in Afghanistan, and later by the Greeks and Romans (the domestic carrot is also called *Daucus carota*). Its root, although white, is rich in carotene (Vitamin B), and no doubt graced many grateful settlers' winter tables before OMAF (Ontario Ministry of Agriculture and Food) gave it a noxious reputation for no better reason than that it "is a common weed in old pastures, roadsides and waste places."

Ditto coltsfoot (*Tussilago farfara*), a very early-flowering plant that grows in wild, wet places, imported by early settlers from Europe probably for its medicinal properties. Hippocrates and Pliny wrote of its beneficial effects on the lungs: inhaling the smoke of its burning leaves was said to relieve croup, and P.H. Gosse, in his rather odd book *The Canadian Naturalist*, published in 1840, noted that it is "useful in catarrhal affections." In French folklore, however, the presence of coltsfoot (called *Pas-d'âne*, or donkey's foot) is taken as a sign of poor land: "*Terre à Taconet, laisse-là où elle est; terre à Renoncule rampante, achète-là si tu peux.*" *Taconet* is Tussilage, or coltsfoot; *Renoncule* is buttercup. "Land with coltsfoot growing, leave it alone; land with buttercups growing, buy it if you can."

Goat's-beard is on the list, and not just common goat's-beard (*Tragopogon pratensis*), but all species of *Tragopogon*. Common goat's-beard is a tall plant (usually one metre, but I've seen it growing head-high in a neglected front yard near Queen and Bathurst) that looks like a thick stalk of grass surmounted by a yellow dandelion flower. It even has a long taproot like a dandelion. The bracts under the composite flowers are pointed and somewhat longer than the flower petals, and the fruiting heads have long beaks tipped with white, feather-like plumes, from which it gets its name. The remarkable thing about goat's-beard is its heliotropism: the yellow flowerheads open early in the morning aimed directly at the rising sun and follow the sun's progress across the sky during the day. Even though the heads close at noon (so reliably that farmers used it to time their lunch breaks), the closed heads continue to follow the sun until it sets. OMAF reports that "it is usually restricted to abandoned fields, roadsides and waste places, but may invade grain fields." Another species of goat's-beard, *Tragopogon porrifolius*, better known as salsify or oyster-plant, differs from common goat's-beard only in that it has purple flowers and its "beard" is brownish, but is actively cultivated for its edible root, which resembles a small parsnip and tastes like oysters; a naturalized European, "Mammoth" salsify seeds are standard fare in most Canadian seed catalogues, even though OMAF says all species of *Tragopogon* must be destroyed, and anyone found disseminating its seeds can be fined.

We have, more or less, lost the war on weeds. My bushel basket full of garden-variety weeds represents a hollow victory after a minor skirmish, because the weeds will be back. The taproots of most of the dandelions I pulled up had snapped off about twelve centimetres below the surface, which means each root will be sending up two dandelions in the next week or so. Their persistence leads me to wonder whether the whole weed-versus-wildflower question isn't one of semantics. Perhaps, in order to win any kind of battle against weeds, we should just stop calling them weeds, the way the Britons stopped calling their oppressors Vikings and started calling them Anglo-Saxons. After all, the plants to which we have assigned the word "weed" have changed as often as fashion and economics have dictated. Maybe fashion and economics are the real enemies, and weeds the poor vessels into which we pour our ire.

Fall

On our Canadian climate I've little to say,
As I've lived in it many years and cold days,
This present month, October, without strife,
Is the beautifullest I ever saw in my life.

– *James Gay, "On Our Climate,"* 1885

TORONTO
AFTER DARK

Some parts of Toronto are only superficially urban. I used
to live in the west end of Toronto, on one of those
streets the houses of which were all crowded close to
the sidewalk, like blocky spectators at a parade. The impression was of shoulder-to-shoulder spacelessness and constriction. The front windows all looked blank – literally blinded –
and the porches were dark, dusty places where kids' bicycles,
stacks of yellowing newspapers and blue boxes were stored.
The trees that lined the street were old, their roots humped up
through the grassless earth like scaly snakes, their leaves
hanging limply in the still afternoons. Many of the front yards
had been paved over and were being used as driveways.

But walk through the house and step into the back yard,
and you entered what I always thought of as "the real
Toronto," for behind the square of houses that formed the
block there were literally acres of trees, flowers, and vegetables, carefully tended, beautifully kept, and clearly the focus
of the lives of the people who lived in the houses. You quickly
realized that the original builders had merely sacrificed their
showy front lawns in order to have this neighbourhood paradise behind. Sunny, open decks and vine-covered patios
graced the rear of the houses, manicured grassy walkways

wove among sun-filled beds of flowers and led to huge vegetable plots clustered around garden seats and dove palaces.
There were fences between each neighbour's yard, but they
were mere tokens, often little more than spindly supports for
scarlet runner beans and trailing morning glories. In the real
Toronto, poor fences make good neighbours. There were fol-
lies and ha-has and gazebos. There were no alleys, no asphalt
driveways with two-car garages, no telephone poles. All
those things belonged to the front world; the back world was
Eden; to it belonged only the things of nature.

Our house was a semi-detached, and the neighbours in the
other half were a retired Italian couple named Joe and Maria.
Joe was a short, stocky man in his sixties, with a bulbous nose,
a thick mound of pure white hair, and fingers like calloused
pepperonis. Maria was taller, though she walked with a stoop,
and her intense eyes and aquiline nose gave her a kind of
tragic, vulturous look that belied her gentle nature. Joe had
come to Canada thirty-five years before, had worked in construction, brought his wife over after fifteen years of separa-
tion, and moved into this house where they raised six
children. He spoke a little English, enough to pass the time.
Maria, though she had lived in Toronto for twenty years,
knew exactly two words of English: "hello," which Joe must
have taught her on the way over, and "church," which she
pronounced "chorch."

Joe and Maria had the best garden on the block. Directly
off the back porch was a framework of galvanized plumbing
pipe that was canopied with grape vines from May to October; a picnic table rested in its shade, and a gravel path led to
the garden, where the rich, black earth practically shot
enough tomatoes, green peppers, green beans, and lettuce
into the air to feed their entire extended family twice over.
The garden beds were watered by a complicated network of
rubber drip hoses connected to a tap that protruded through
the grass at the end of the arbour; every summer Joe would sit

at the table in the shade of his grape vines and watch with satisfaction the inevitable ripening of his labours.

My own garden was a sorrier affair. The previous tenants had more or less let it run its course, and I had a lot of restoration to do. A few parched tomato plants and some yellowing spinach fought a rear-guard action with the creeping Charlie and were successful enough to allow the zucchini full reign; it tended to spread out over the end of the yard, climb the fence, and was threatening to invade Joe's determinedly unzucchinied domain.

One Sunday afternoon in September, Joe and I were sitting on our respective sides of the fence, gazing at our contrasting gardens, absorbed in our own thoughts. Maria had said hello and gone to chorch, and Joe had poured us two huge tumblers of his home-made wine from a twist-cap Pepsi bottle. It was a functional wine, tantalizingly red, with a full-bodied fizz that tickled the back of the throat and occasionally brought tears to the eye. Joe was very proud of it. I felt honoured to be sharing it. As we drank, he asked me the names of his plants in English.

"Tomato," I said.

"Ah," he said. "Toe-may-do. *Pomodoro*."

"Lettuce."

"Lead-us. *Lattuga*," he said, "milk-weed."

He pointed to my luxuriant zucchini as a healthy man might point to a patch of scabies. "How do you call that?" he asked, making a face.

"Zucchini," I said, surprised. "Isn't that Italian? What do you call it?"

"In Italy we call it pig squash."

It was a nice afternoon, but I was about to ruin it. A few days before, we had discovered cockroaches in our house. At first there had been one or two on the kitchen counter, then a few more near the sink, and that morning I had rolled the refrigerator away from the wall, taken off the rear panel, and . . .

well, let's just say that Joe's wine was even more welcome than usual that afternoon. Cockroach infestations should not come as a horrendous surprise to anyone living in downtown Toronto, but this one put me in a delicate situation, because I knew that if we had them on our side of the house, then Joe had them on his, although he may not want to admit it; and if we had our side fumigated, the cockroaches would merely decamp over to Joe's side until the fumes subsided and then return to their former haunts. I had to convince Joe to have his side fumigated at the same time ours was, and I didn't know how he would take the suggestion.

"Joe," I said, "I've got some bad news. We have cockroaches."

"Eh?" said Joe. "Cockaroachis, what is that?"

"Little insects," I said. "Bugs. In the house. All over. We've got to get rid of them."

"Little bugs?" he said. "They crawl like this?" His hand made a very cockroach-like scuttle across his picnic table. I nodded. "Yeah," he said, "I gottem too. I never had before. Where they come from?"

"I dunno," I said. "Maybe we brought them home in a box from the grocery store, or the beer store. They hide inside the cardboard. Maybe they were here all along. Who knows? But we have to call the fumigators to come and spray our house. I think you should have your house done too."

"Fumigators, *fumigazioni*," he said, reaching over the fence to refill my glass. "Yeah, you call them. I would have called myself, but I didn't know the word in English: cockaroachis. Is a funny word. What it means?"

It doesn't really mean anything. Both the male and female of the species is a "cock," though neither is a fowl, and a roach is a small freshwater fish of the carp family. Our word "cockroach" is a corruption of the Spanish *cucaracha* and was first used in English by Captain John Smith, who in 1624, in his report of the Virginia colony, mentioned "a certain India Bug, called by the Spaniards a cacarootch, the which creeping into Chests they eat and defile with their ill-scented dung," which puts the lie to the contention that cockroaches first appeared in the United States in 1846, with the building of New York's Croton Dam – German cockroaches were called Croton bugs or water bugs in New York in the belief that they were brought over by German immigrants who came to work on the dam. The truth is, *cucarachas* have been around, even in the New World, probably forever: Cadwallader Colden, a Scotch-American physician, botanist, mathematician and the lieutenant-governor of New York from 1761 to 1766 (he also wrote a *History of the Five Indian Nations of Canada* in 1727), was of the belief that cockroaches – which he called "millbeetles," although cockroaches do not mill and are no more beetles than they are fish – were brought to North America from the West Indies on trade ships. Peter Kalm, however, that itinerant Swedish naturalist, a student of Carl Linnaeus who travelled throughout North America from 1747 to 1751, disagreed with Colden: "I have reason to believe that these insects have been in North America since time immemorial." At any rate, in 1750 Kalm found cockroaches "in almost every house in the city of New York," although he said there were none at all in Canada, and the French people he met in Quebec had no name for the insect (they still don't: although in France it is called *cafard*, in Quebec it was and is known simply as *coquerelle*, an obvious borrowing from the English). The Dutch in New York, Kalm noted, gave it the name *kackerlack*, which is either another version of *cucaracha* or else the sound the insect made as it scurried across the bread board whenever anyone went into the kitchen carrying a candle.

A story in the Toronto *Star* in April 1985, reporting that "90 percent of Toronto's apartment buildings have the battle-hardened insects," included a drawing of an American cockroach (*Periplaneta americana*) with the caption: "Behold the enemy," but in fact there have been only isolated cases of American cockroaches in Toronto. What we have in Toronto is the German cockroach, *Blattella germanica*, slightly lighter in colour and much smaller than its American cousin (about 2.5 cm, or half as long). All cockroaches (and there are more than 6,000 species worldwide, with, entomologists suspect, at least another 5,000 species waiting in the woodwork to be described) are long-legged, long-antennaed, somewhat flattened insects belonging to the family Blattidae, of the order Orthoptera, which means "straight-winged" and includes such insects as walking sticks, mantids, crickets, and grasshoppers. All but a very few species are strictly tropical, and only 1 percent are considered pests. Of the fifty-five species found in North America, so far only the German cockroach is commonly found in Toronto – although the Asian cockroach has been recently discovered in Florida and is reportedly moving north. This would make for interesting times: the Asian cockroach can fly, is attracted to light, and can mate with German cockroaches to form what a spokesperson at one Toronto pest-control company says would be "a super-roach." The American and German cockroaches have become so accustomed to sharing human habitations that neither is found any longer in the wild; both are introduced species, expand their range by hopping onto human transportation systems, such as corrugated cardboard boxes, and don't seem to bother spreading very far inland from their points of entry.

Cockroaches evoke as explosive a reaction in humans as humans evoke in cockroaches. Turn on the lights in a kitchen that has cockroaches, and watch both humans and cockroaches dive for cover. There is probably a psychological term for the fear of cockroaches: blattophobia? Sue Hubbell, author of *Broadsides from the Other Orders: A Book of Bugs*, notes that there is a name for the fear of crane flies – tipulophobia – so why not one for cockroaches? German cockroaches have an oily coating on their cuticle that enables them to scuttle into the tiniest of cracks; humans, I can attest, tend to flee into open spaces, like back yards. If we'd been living in a high-rise apartment, I might be dead now. In 1985, two researchers at Toronto's Department of Public Health actually conducted a telephone survey of attitudes and responses to cockroach infestations in apartment buildings. The survey included such questions as: "Would you say your apartment building has a cockroach problem?" Fifty percent of respondents said yes. "How serious would you say the problem is?" Forty-four percent said serious, sixteen percent said very serious. To no one's surprise, the researchers found that 89 percent of the respondents considered cockroaches a health hazard, 91 percent thought they were "ugly to look at," and 94 percent found them "a source of anxiety." No kidding. A few years ago, when two people moved out of their apartment in Mississauga because of a severe cockroach infestation, a Peel County judge awarded the vacated tenants a $434 rent rebate, back to the time they first notified their landlord of the problem. In his nine-page ruling, Judge Martin Morrissey said the landlord was responsible for pest control, and that the presence of cockroaches "had a severe emotional and psychological impact on the tenants."

Half the apartment dwellers in Toronto, however, live with that severe emotional and psychological impact every day, and the results can be serious. Consider the case of Mirza and Parveen Baig, whose seven-month-old son Majid died three months after their Scarborough townhouse was sprayed for cockroaches in January 1982: an autopsy showed that the child's liver "might have been damaged" by exposure to diazinon, the insecticide found in most over-the-counter anti-cockroach products. The Baigs' landlord had sprayed their basement and first floor while the Baigs were still living in the house and without first removing the family's food and clothing from the cupboards: the baby's cereal was later found to contain 4.4 parts per million of diazinon. Unfortunately, we have only begun to tally the long-term cost of our losing pesticide war against insects.

Quite apart from the question of toxic chemicals, many people suffer from the side-effects of having cockroaches without being aware of it. Cockroaches are more than just aesthetic nuisances; because they spend a lot of time around putrefying food, they carry pathogens that cause diseases in humans. Cockroach guts have been found to carry fifty-six different types of bacteria, at least fourteen of which are known to be pathogenic to us, including staphylococcus, streptococcus, and coliform bacteria such as *E. coli*. Cockroaches have also been found to be reservoirs of salmonella in India; because certain species have developed resistances to chemicals, the salmonella bacteria can hide, as it were, in a cockroach until the treatment is over, and then re-emerge to infect humans again. Studies also show that humans susceptible to allergies, and especially those with asthma, react strongly to dust containing particles of cockroach skin after moulting; in fact, studies at the University of Kentucky suggest that more than half of the 15 million North Americans who suffer from asthma are also allergic to cockroaches. Of all insects assessed as vectors for human diseases, cockroaches come second behind houseflies.

North Americans spend more than $350 million on cockroach poisons every year, and nearly $1 billion on research into new roachicides. The list of trade names sounds like the kind of computer games we try to keep our kids away from – Maxforce, Combat, Demon – and their chemical constituents are more or less what would be left in the Great Lakes if you removed the water: hydramethylnon, dioxacarb, disodium octaborate tetrahydrate, cyfluthrin, cypermethrin, chlorpyrifos, propoxur, bendiocarb. Except they aren't just floating around in Lake Ontario; they're in the air we breathe every day. Combine our skirmishes with roaches with our blitzkrieg on agricultural pests, and we begin to wonder when scientists will have to measure oxygen instead of DDT in parts per million. And it is, after all, a losing battle. In the past two decades, our use of agricultural pesticides alone has increased by 800 percent, and over the same period our loss of crops to insects has jumped from 7 percent of total production to 13 percent. Faced with such incontrovertible evidence of failure, some scientists are finally beginning to suspect that bombarding insects with an ever-expanding chemical arsenal might be the wrong approach.

The problem is that insects very quickly develop immunities to things that are dangerous to them. Insects have been around a long time. There were insects on Earth long before there were dinosaurs: the Devonian Period, 600 million years ago, is known as the Age of Insects (much to the annoyance of fish, which also showed up during the Devonian), and the cockroach is thought to have originated then, although no fossils exist from that period: the snub-headed, flat-bodied roach *Eoblattina temporis* was found in an Upper Carboniferous deposit (250 million years old) in Commentry, France, and looks uncannily as though it had been squished under a rolled-up newspaper yesterday rather than trapped in limestone sediment 250 million years before Raid. Of the eleven orders of insects certain to have originated during the Carboniferous, only the cockroach is still with us in anything like its original form: by all evidence, it is virtually unchanged (except for its resistance to certain chemicals) from its first appearance in the fossil record. A reconstructed Carboniferous cockroach, if such a thing were to exist, would be almost indistinguishable from a cockroach plucked from an apartment block in Downsview yesterday, and that is something that cannot be said about any other organism on the planet with the possible exception of one or two types of algae.

Perhaps because of this longevity, the cockroach doesn't seem to be fazed by the latest chemical roachicides, at least not for very long. What for relatively young insects, such as

bees or butterflies, would be a devastating threat to their continued existence is for the cockroach a mere bagatelle in the
yawning stretches of time it has been on this Earth. Cockroaches, like mosquitoes (which also disappear off the evolutionary time chart), seem to develop immunities to chemicals
as fast as we can come up with them. Diazinon, bendiocarb,
propoxur, boric acid – individual communities of cockroaches
have been found that have developed immunities to all of
them; even the newest and most hopeful family of insecticides, the so-called amidinohydrazones, touted as having the
potential to reduce New York City's roach population by 50
percent in three years, is expected to be impotent within a
decade. Reports of new pesticides that will virtually wipe out
the cockroach problem in our major urban centres are beginning to wear a bit thin, like the repeated promises of politicians that a new series of spending cuts will go a long way to
eradicating the federal deficit. "The chemistry is there," Dr.
Austin Frishman, an entomologist and pest control consultant
told the New York *Times* a few years back, "to keep roaches
under control for the next ten years if we play our cards right."
Ten years? What's ten years in the 250-million-year span of
the cockroach? I have a feeling they can wait us out.

The changelessness of cockroaches over such vast
stretches of deep time does make them interesting subjects to
scientists investigating how genetic changes drive the evolutionary process, however. Randall Alford, a forest entomologist at the University of Maine, has written that "the cockroach is the most frequently used insect for the study of behaviour, anatomy, and physiology" in the lab, partly because
cockroach colonies, or cultures, are easy to raise (a damp
cloth and a handful of rat food will keep a garbage pail full of
them alive and reproducing for months), but also because
cockroaches have remained primitive organisms against
which evolutionary change in other organisms can be mea-
sured. "Most other insects," Alford writes, "have experienced

an evolutionary reduction in features to provide more special-
ization. Cockroaches, however, are generalists, eating almost
anything and, through the ages, they have been successful
with life as they define it. . . . There have been no incentives
or pressures to change." Geneticists, molecular biologists, and
immunologists, then, can subject cockroaches to various artificial stimuli to determine which ones produce the specific genetic or behavioural changes that have been detected in other
insects. It's a bit like finding a perfectly preserved Neanderthal
and finding out what, if anything, it would take to turn him
into a defensive lineman for the Hamilton Tiger-Cats.

In the course of studying cockroaches either for pure science or, more often, to be able to devise better ways to kill
them ("Know thine enemy"), scientists have learned a lot
about them – especially about German cockroaches. Attempts to discover the secret elixir by which female cockroaches attract male sexual partners – with a view towards
putting some of this elixir in a poisonous trap and luring great
hordes of panting male cockroaches to certain death – have
produced some fairly intimate glimpses into the private lives
of cockroaches. The elixir, otherwise known as a pheromone,
is a scent that is emitted from a gland under the cockroach's
wings; it is detected only by members of the same species, and
in the case of German cockroaches only by members of the
opposite sex, and it drives them crazy. When the female is
ready to mate, she raises her wings and fans her pheromones
towards the male object of her attention, who immediately
approaches her, comes to rest with his forehead very close to
her own, and the two then engage in a kind of mutual antenna
massage, each partner gently stroking the other's antennae
with his or her own. When he can stand it no longer, the male
makes a quick, 180-degree turn, raises his own wings and fans
a whiff or two of his own pheromones back towards the female. This excites the female even further than the antenna-
massage, and she climbs up on the male's back in order to get

her mandibles closer to the source of his elixir. The male then wiggles backward under her until he is in the correct position for him to hook onto her exposed genitalia with his own barbed equipment, whereupon he swings about in another 180-degree arc so that the two are standing tail-to-tail, tethered together by his grappling aideagus like a pair of mating Wiener schnitzels. They remain thus locked in what some unromantic scientists refer to as the "sperm transfer position" for up to two full hours. If cockroaches could smile, they'd be smiling now.

Throughout this barbed embrace, the female German cockroach gathers and stores her partner's sperm in a special sac known as the spermatheca and uses it to fertilize her eggs, which in their turn are stored in a special sac known as the ootheca. Whereas the females of most other cockroach species drop their oothecae as soon as the eggs in it are fertilized – some species construct elaborate casings for them out of sand and saliva, others merely deposit them in piles of leaf mould or compost where they will remain warm and moist until the young come out, twenty-eight days later – the female German cockroach keeps her ootheca in her abdomen until the day before the eggs are ready to hatch. Some researchers think this is an adaptation to living in human habitations, since dropping oothecae on floors and counter tops, where they would either desiccate or be swept up by us (you bet they would), would not be a particularly successful reproductive strategy, but it is also possible that German cockroaches have become as ubiquitous as they are in our houses because they have always had this mammal-like propensity to nurture their eggs inside their bodies until the young are ready to hatch.

Each ootheca contains thirty-five to forty nymphs, each of which goes through a series of seven moults, or instars, until it reaches the adult stage, a process that takes about 103 days, depending on the ambient temperature (the warmer it is, the faster they mature). An adult female can live another 200 days, producing four or five oothecae in that time. Rough math arrives at a large number of cockroaches very quickly: it has been calculated that five fertile female cockroaches, left to their own devices and desires, can produce 45 million descendants in one year.

They are not, therefore, left to their own devices and desires. We have devised a huge array of toxins aimed specifically at cockroaches: one-quarter of every dollar spent on insecticides in North America is spent on anti-cockroach weaponry. Some of it is effective, much of it isn't. The *Common Sense Pest Control Quarterly*, a publication originating in Berkeley, California, recommends habitat control over chemical spray-bombing, but adds that if, after eliminating places where roaches can hide or migrate, you still want to use a chemical control, then you should look for materials that are not also harmful to other species, including humans. Boric acid is very effective and also safe – it is used in contact-lens cleaners – but is rarely used by professional pest control companies because it is applied as a dust and is therefore more time-consuming to apply than aerosol treatments. Diatomaceous earth is also safe and effective. Diatomaceous earth is a dust made from the crushed shells of fossilized unicellular organisms (diatoms), a kind of pre-limestone powder, and kills cockroaches and other insects by cutting through their tough outer carapaces when they crawl over it: an insect with a cut shell dehydrates rapidly. Boric acid and diatomaceous earth take a few weeks to work, another reason pest control companies don't like them, but at least they won't kill anything but the cockroaches.

If you have cockroaches and don't like chemicals, you could always get yourself a tokay gecko. A few years ago, I read that New Yorkers were buying tokay geckos from pet shops, because tokay geckos were reputed to eat cockroaches. Geckos are small lizards native to southeast Asia, with clinging pads on their feet that allow them to climb up walls and even across ceilings. The tokay of Malaya is one of the most easily bred members of the family; for $20, according to the article, apartment dwellers could buy a five-to-twenty-centimetre-long tokay gecko which, left to roam the apartment, would gobble up to 200 cockroaches a night: "all the owner has to do is leave out a bowl of water."

Intrigued by this report, I called my friend John Acorn, an entomologist who keeps geckos in a large glass tank in his house in Edmonton, and asked him what he knew about tokays. "The tokay gecko," he replied, "or *Gekko gecko*, is the largest of all living geckos. It is attractively marked with orange and blue spots on a velvety grey background. I've heard that in some places the arrival of a tokay in your home after you get married is a sure sign that all is well, and that you will have plenty of kids. But as for keeping them as pets, I'm not so sure; the truth is, they can be bad-tempered, hard-bitin' hombres.

"A few years ago," he continued, "I was invited over to a friend's house to see his basement, which he had set up for reptile rearing. He had hundreds of snakes and lizards in one half of the basement, and in the other half he was raising the food: dozens of mouse and rat cages, culture jars full of rotten bananas and yeast, and swarms of fruit flies everywhere. The whole basement was sweltering and humid in the extreme, as appropriate in a tropical jungle. At one point, my tour was interrupted by a bellowing bark: 'Geck-oh! Geck-OH! GECK-OH!'

"'A tokay gecko,' I remarked.

"'Want to see it?' he asked.

"'No thanks,' I said. 'I've seen them before. I just hadn't heard one.'

"'Well,' he said, 'you haven't seen one like this.' And, taking up a pair of long, wooden tongs, he grabbed a young mouse from a cage and dangled it beside a stack of snake cages set against one wall. In a few seconds, a huge tokay gecko, about fourteen inches long and with a forehead at least two and a half inches across, shot out from behind the cages, grabbed the mouse from the tongs, killed it with one crushing bite, and swallowed it whole. Ever since then I've been of the opinion that keeping a tokay for roach control might be an example of overkill: it's sort of like having a nocturnal, bald, rabid, anti-social Yorkshire terrier that can walk up walls."

An easier solution might be to put all your belongings in a large plastic bag and blow it up with carbon dioxide. That's what Chris Christiansen, a biologist at the University of Kentucky, did: he piled clothing and furniture from several roach-infested buildings into a huge, 6-mil polyethylene bag, sucked all the air out of it with a vacuum cleaner, then filled the bag with carbon dioxide. Then he sealed the bag and left it overnight: in the morning, he reported, the cockroaches were "either dead or extremely sluggish," and the sluggish ones died within a few hours.

Laboratory scientists are coming up with some pretty ingenious methods of their own, by using the cockroach's own biology against it. When biologists discovered that the female cockroach's sex pheromone is triggered by decreased levels of juvenile hormone, for example, they were able to manufacture insect growth regulators (IGRs) by synthesizing juvenile hormones and pumping them into female cockroaches, keeping them trapped in the adolescent stage and therefore unable to reproduce. They grow into adults, but they have deformed wings and genitalia. Other research isolated the cockroach gene responsible for the growth of chitin, the material that forms the insect's outer shell. Interfere with the chitin gene, and when the cockroach sheds a skin during one of its seven moults, it doesn't grow another one.

Biological warfare has also spread to planting parasites. The fungus *Metarhizium anisopliae* was used against the Oriental

cockroach in Czarist Russia and has been introduced here
against the German cockroach with interesting results. The
fungus attaches itself to the insect's exoskeleton and eats away
at it, sending its roots down into the insect's gut and consum-
ing it from within. The process takes about a week, but it's a
satisfying week. Then there are nematodes. Nematodes are
microscopic worms that crawl up the cockroach's anus and
deposit a bacteria in its digestive tract that kills the cockroach; the nematodes then eat the carcass. Life at the microscopic level can be just as grim as the macroscopic kind.

Joe and I didn't know any of this, of course, when we shared a
bottle of wine under the grape arbour in his back yard. We
didn't know anything about nematodes or chitin inhibitors or
tokay geckos. All we knew was that when we turned on the
light in our kitchens at night, the corners of our eyes would
twitch and our legs would fill up with adrenaline. It would be
a long time before I would bring myself to remove the panel
from the back of a refrigerator again. If I knew then what I
know now, I wouldn't have called in the chemical army. I
would have tried to adopt the attitude expressed by the
phrase "When you're dealt a lemon, make lemonade."
    This seems to have been the attitude of Catherine the
Great, who, in a Russia overrun by Oriental cockroaches, became a great fan of cockroach racing. Russian officers exiled to
Constantinople after the October Revolution in 1917 also enjoyed cockroach racing. It could catch on here. You dim the
lights, place a handful of cockroaches – each painted with a
number and its own racing stripe – at the end of a long runway,
then quickly turn on the lights. Place your bets; entomology
can be fun and profitable. The sport has, apparently, been revived in the cash-poor former Soviet Union and is a popular
attraction at the St. Petersburg (formerly Leningrad) Circus.
Nikolai Popov, one of the organizers of the event, has said that
his "first priority is to revive this ancient Russian tradition," but

it is clear that his second priority is to make a lot of money doing it. He plans to sell satellite TV rights, institute off-track betting by telephone and, the London *Daily Telegraph* reports, he is applying to the International Olympic Committee to have cockroach racing introduced as an Olympic sport. It seems a come-down from the days of Nadia Comaneci and Red Army hockey, but times have changed.

PIGEONS ON
THE GRASS, ALAS

N ear the corner of Lakeshore Boulevard and Royal
York Road there is a tall, narrow apartment building
that is instructive in the ways of pigeons. Almost
every balcony on it is adorned with some weird device: plastic
flowers with white centres and madly spinning red petals; a
desperately paddling Daffy Duck trying to outswim a mania-
cal Elmer Fudd; a smiling Easter chick with a polka-dot bonnet and a basket that flashes like a mirror in the sunlight. One
balcony has a wire strung across it with thirty-two white rags
tied to it, each one flapping in the fierce wind generated by
the building itself; another has a wire similarly strung and a
row of red plastic bags; yet another has a huge plastic owl of
no discernible species, its feet screwed to the railing and its
head turned permanently to the rising sun. All these devices
express a single desire – to discourage pigeons from landing
on the balconies – and none of them work. In an hour's watching from a parking lot across the street one October morning,
I counted a flock of sixteen pigeons wheeling about and landing on various parts of the building; out of a total of 211 landings during that one hour, eighty-seven were on balconies
displaying something that was supposed to say Shoo; forty-
three were on unadorned balconies; and eighty-one were

somewhere along the edge of the building's flat roof, usually downwind of the ventilator shaft. Of all the balconies, the one most preferred by the pigeons – twenty-three landings – was the one with the fake owl screwed to the railing. They seemed to like the company. Pigeons don't appear to demand a lot from their friends except that they keep still and be quiet.

It shouldn't surprise anyone that, at least in this part of Toronto – the stretch of lakeshore known as Mimico – pigeons are hard to control, for the word "mimico," in the language of the Mississauga Indians, means "the place where pigeons gather." Of course, the Mississaugas weren't referring to the same kind of pigeons I watched from my car on that cold day in October. I was counting our ordinary, garden-variety pigeons, known to biologists as rock doves, *Columba livia*. The Mississaugas were talking about passenger pigeons, *Ectopistes migratorius*, which were native to North America. The residents of Mimico no longer have to worry about passenger pigeons: the last Torontonian to see one was J. Hughes Samuel, who reported a flock of five on Centre Island on July 6, 1900, and no one has seen a live one at all since September 1914, when Martha, the world's last living passenger pigeon, died in a cage in the Cincinnati Zoo.

Passenger pigeons differed from rock doves in several ways. They were much more colourful, for one thing. Rock doves might as well have been named for their colour, because their dominant hue is a kind of granite or slate grey, with white underwings and a slightly greenish-red, iridescent sheen about the head and neck, like an oil slick. The passenger pigeon, on the other hand, was a comparative palette of hues: its throat and chest were a dusky, damasky rose, the colour of a sunset; its head, shoulders, and back were a bluish grey; its long, forked tail was grey-red, and there were watery-green stripes across the back of its neck. Only its underside, from the legs back along the tail, was white. The other differences are mainly behavioural. *Columba livia* doesn't migrate, which is why I could watch a flock of them in Mimico in October; *E. migratoria*, as its name implies, was a wideranging bird, migrating in almost unbelievable numbers from as far north as Hudson's Bay and the Mackenzie Valley down to the Gulf of Mexico and back. Even its English name is derived from this wandering trait – the English word "passenger" comes from the French *passagère*, meaning a voyager (rather than someone just along for the ride), and a migratory bird is *un oiseau passager*. The passenger pigeon was the only dove in North America that migrated.

There is a single male passenger pigeon, stuffed, on permanent display at the Royal Ontario Museum; it is perched demurely on a branch in a recessed compartment in a wall, behind glass, as though caught in a freeze-frame on a television screen. In the 1930s, ROM possessed a larger diorama containing a dozen or more passenger pigeons prepared by its pre-eminent bird artist Terry Shortt: passenger pigeons landing, soaring, courting, with great flocks of them painted on the backdrop. That collection has disappeared along with the passenger pigeon itself. There is another stuffed male passenger pigeon in a glass case in the Haliburton Highlands Museum, in Haliburton. In front of the display, there is a notice that reads:

Passenger Pigeon

became extinct in 1914. The last bird (a 29-year-old female) died at the Cincinnati Zoo. Since they flew in flocks of literally millions it was easy for people to slaughter them thoughtlessly, taking enormous quantities at a time.

These birds roosting and nesting in the trees (sometimes as many as 100 per tree) caused great destruction to large areas of forest.

There are many incredible stories describing the brutal fate of the flocks of passenger pigeons.

All of which is true. When the passenger pigeon migrated, it did so "in numbers that seem to have exceeded those of any other bird of which we have record," writes Margaret H. Mitchell in *The Passenger Pigeon in Ontario*, published in 1935. Flights of migrating pigeons would darken the skies for days – Major W. Ross King, writing in 1866, records a single flock of pigeons passing over his quarters at Fort Mississauga, near Niagara, for three solid days, blocking the sun and filling the air with the thunder of their wings. Alexander Wilson, the great nineteenth-century American ornithologist, estimated a flock he witnessed to have contained 2,230,272,000 birds (he was a fast estimator). Nesting colonies of several million birds covering hundreds of square miles were common in Canada throughout the 1800s. In Huron County, near the headwaters of the Maitland River, a nesting colony completely covered an area eleven miles long and thirteen miles wide. There were two such sites in the Toronto area, one in Mimico (hence its name) and another in a large stand of pines a few miles east of Yonge Street, in Aurora, known as Pine Orchard. "To say that there were millions of nests there," wrote N. Pearson, responding to a ROM questionnaire in 1926, "would be a mere assertion." As the Haliburton notice says, there were often up to 100 nests in a single tree; the weight of nesting birds would bring large branches and even entire trees crashing to the ground. "The mind is lost in endeavouring to form an idea commensurate with these vast numbers," wrote P.H. Gosse in *The Canadian Naturalist* in 1840, "and this small and apparently insignificant bird may justly be considered one of the wonders of this western world."

Unfortunately, this wonder of the western world was also a nuisance to farmers. In the fall, passenger pigeons lived mainly on nuts – stands of beech trees were literally buried under pigeons as they passed – but in the spring and summer they devoured huge quantities of berries, tree seeds, and grain. Farmers would seed their fields with corn, only to have pigeons pick them clean in an hour. And when it was discovered that passenger pigeons were excellent eating – in Quebec, *tourtière* was originally made from passenger pigeon (the name of the meat pie comes from the French *tourte*, or pigeon, from which we also get the phrase "turtle dove") – the bird's fate was sealed. "In the season," continued N. Pearson, "Pine Orchard was inundated with people from the country about. Waggon loads of Farmers with their sons from miles about came during the daytime and at night with lanterns and torches and slaughtered with wholesale vigour. Caught the old birds and wrung their necks and carried off the squabs in bags by the waggon load."

During the 1860s, pigeons were regularly shot by Sunday hunters in Toronto, mostly in a large tract of open ground south of Bloor between Bayview Avenue and Parliament Street, where the Rosedale Valley Road runs below St. James Cemetery, the site of John Graves Simcoe's summer residence, which he named Castlefrank for his son, Francis. During migration, so many "pigeoners" haunted the Castlefrank area that for many years it was known as "Pigeon Green."

Pigeons harvested in the outlying farming districts were shipped into the city and sold at the St. Lawrence Market for five cents apiece, or to hotels and restaurants for ten cents each or a dollar a dozen. "We usually hung the birds in small lots of two or three overnight, to cool off," wrote M. W. Althouse, a farmer in Middlesex County, responding to the same ROM questionnaire, "then packed them in layers of straw in apple-barrels, and sent them . . . by express. My father's net sometimes furnished three or four barrels in a single day," with each barrel holding "about 100 or 120 birds. Sometimes the netting season lasted two or three weeks."

Another market for live pigeons was the Toronto Gun Club, formed in 1871, where pigeon-shooting, or "trap-shooting," had become all the rage by the end of the century. Trap-shooting was so-called because the live birds were held in galvanized iron traps, about the size of bread boxes, with lids that could be opened from a distance by means of a string; when the lid was opened, the floor of the trap swung up and pushed the pigeon into the air, where it was shot by a doughty sportsman standing point blank in front of it. Birds kept in closed iron cages in mid-summer tended to become somewhat listless, it appears, and so measures were taken to ensure a lively target. These included tearing a patch of feathers off the pigeon's back and rubbing the exposed raw flesh with essence of cayenne. Only the Gun Club could afford to import wild pigeons for its spring shoots – they bought them from a dealer in Buffalo. Other Toronto clubs re-sorted to what was no doubt viewed as a come-down in the fine sport of pigeon shooting: the use of domestic pigeons raised for the purpose. Those who used domestic pigeons tried to maintain that they gave better sport, because they were in better physical condition, but everyone knew the real reason was that they were cheaper. By the turn of the century there was another, more compelling reason to switch to domestic birds: they were infinitely more plentiful. The last huge flock of passenger pigeons flew over Toronto in 1878, after which their number steadily declined until, just after the turn of the century, it reached zero. It had been hunted to extinction.

From 1914 to 1925, the Royal Ontario Museum and several ornithological institutions offered cash rewards to anyone reporting a verifiable sighting of *E. migratoria*, and many such reports came in – all of which turned out to be sightings of the similarly coloured mourning dove, *Zenaida macroura*, which are still quite numerous and in fact, in Toronto, are becoming more so. Experts are divided as to whether the increase in mourning doves represents a new, northerly expansion of the bird's historic range, or whether there are simply more mourning doves in the world, but there is no doubt that the hollow, almost loon-like hooting of mourning doves is a sound we hear more and more frequently around our win-dows and bird feeders. Driving through the countryside near Toronto this past year, I have seen more and more mourning doves perched on telephone wires and fences, and friends with feeders say they have a dozen or more regular visitors now instead of one or two, as in the past.

The mourning dove was originally called the Carolina turtle dove until around 1840, when someone in Boston decided its sombre grey feathers made it look as though it were wearing a mourning cloak (has no one but me noticed a shimmer of pale pink about the bird's head and neck?). But its name probably derives more from its song than from its plumage. No less an authority than Percy Taverner, in his *Birds of Eastern Canada*, notes that "its long-drawn mournful note of *'oh-woe-woe-woe'* . . . has given the name to the species. It has a peculiar quality like that produced by blowing softly through the neck of an empty bottle." And what could be more mournful than the sound produced by an empty bottle?

Mourning doves are more solitary than their feral cousins – in fact, compared to pigeons, mourning doves are positively hermitic. Two nests are very rarely found together. A nest may contain three or four eggs twice a year, but this can be deceptive because a female mourning dove will often lay an egg or two in another mourning dove's nest – a practice known as nest parasitism and also found in brown-headed cowbirds and European cuckoos, except the mourning dove is parasitizing its own species. Like the extinct passenger pigeon, mourning doves feed principally on mast crops – beechnuts, acorns, and other soft tree-fruit – but are not above settling down on our bird feeder and beating the squirrels and blue jays to the sunflower seeds. Their strut seems less cocky than that of ordinary pigeons, more low-slung, and their longer tails and thinner necks give them a distinctly dove-like profile when they sit on their solitary perches. They look to me more like love birds – turtle doves – than pigeons.

Certainly no epithet containing the word "love" is likely to be used to describe the domestic pigeon – even the gentler and more scientific "rock dove" seems to stick in people's craws. In fact, we might as well dispense with the word "domestic" as well. It is true that originally rock doves were introduced to North America as domestic pigeons – in Canada, by the way, by Samuel de Champlain, whose colonists at Port Royal brought them over instead of chickens in 1606, as easily transportable sources of food. Marc Lescarbot's *History of New France*, published in 1611, noted that eagles were so numerous around the Nova Scotia colony "that they often ate our pigeons, and we had to keep a sharp lookout for them." Pigeons were later imported by colonists in New England as well – the records of the Council of Virginia for December 5, 1621, mention "Pidgeons and other commodities" among the effects of the settlement.

Eventually, inevitably, these domestic pigeons, of which there were more than a dozen different breeds, escaped, interbred, and became feral, although few actually strayed very far from human habitations. In all of North America there are only four places where feral pigeons live in the wild – in British Columbia's Okanagan Valley, where there are rock cliffs suitable for colonization by rock doves, and in similar nesting sites in Colorado, Wyoming, and Utah. The rest of the continent's millions of *Columba livia* can be found on the ledges, balconies, and railway bridges of its major cities. Toronto's portion seems to be about 20,000.

Sibelius Park, a small, green oasis in the Annex north of Bloor, is one of the few parks in the city with a statue, which may be why it is so popular with pigeons. The bust of the Finnish composer, presented to the city by Finland in 1959, stands on a pedestal at the north end of the park, looking somewhat resignedly down on the pigeons that congregate at its base, many of which seem to have expressed their opinion of *The Swan of Tuonela* in the most graphic manner available to them. The word "pigeon" comes from the French *pijon*, which in turn is from the Latin *pipire*, meaning "to whimper," but in this quiet, sun-filled corner of the city it is the guano-bestrewn maestro who seems to be whimpering while the pigeons strut contentedly about in their peculiar jerk-necked fashion, pecking at the bread crumbs and soft seeds broadcast for them daily by the park's human frequenters.

Like their wilder cousins, pigeons prefer nuts and grain, but have adjusted to city life by managing to garner nourishment from just about anything, even white bread. Unlike mourning doves, pigeons have feet that are not made for gripping; they cannot roost on wires, small branches, or the thin edges of metal strips – they were originally cliff dwellers and still prefer to walk around on the parts of buildings that most resemble ledges. Their nests are similarly throw-backs to their cliff-dwelling days – loose twigs kicked together on flat surfaces, the primary purpose of which seems to be to prevent their somewhat roundish eggs from rolling over the edge rather than to provide any sort of warmth or comfort. Unlike other, perhaps tidier birds, pigeons do not catch the fecal sacs of their young with their beaks and toss them over the side, but let the guano drop into the nest to mix with the twigs and feathers and food remnants to form a more cohesive glob of nesting material.

Pigeons are thus favoured hosts for many forms of ectoparasites – including the usual ticks, fleas, bedbugs, louse flies, and kissing bugs – and some fairly serious diseases. One of these, a fungal disease known as cryptococcosis, can be passed on to humans who are exposed to the spores in dried pigeon dung and can lead to a form of meningitis.

This is often used as an excuse to launch massive and expensive campaigns to rid cities of pigeons. "Feathered rats" is an often-heard and entirely undeserved epithet. According to Sandra Zelmer, in Toronto's Animal Control Office, the city receives as many complaints about pigeons as it does about raccoons, skunks, and squirrels. "We're only mandated to act on raccoons and skunks," she says, "but there are periodic attempts to add pigeons to the list. Every now and then someone gets a bee in their bonnet about pigeons, but they always come to nothing." Among the suggestions for pigeon eradication was a plan to feed them grain laced with a chemical sterilant that has been successfully used in Paris and some American cities, but opponents argued that the plan put other grain-feeders at risk as well. In 1988, the city of Etobicoke (which includes Mimico) announced a plan to spend $20,825 feeding strychnine-laced corn to 500 pigeons that had taken up residence under the railroad overpass where Dundas Street West crosses Royal York Road, because their droppings were considered hazardous to pedestrians. It took the combined efforts of the Toronto and Ontario Humane Societies and the Etobicoke Animal Welfare Association to convince the city authorities that, as Liz White of the THS put it, "pigeons are sophisticated enough to feel pain and suffering." So the city switched to a plan to trap the pigeons instead and remove them to another locality. Traps costing $940 each were placed under the overpass and were almost immediately stolen or vandalized. In the end, plastic netting was strung under the bridge to keep the pigeons off the girders. Tattered remnants of this net can still be seen poking out of pigeon nests in some of the more isolated corners of the bridge.

Pigeons lay two off-white eggs at a clutch, forty minutes apart. Wild rock doves do this twice a year, like most wild birds – one clutch in the spring, another in early summer. City pigeons, which do not migrate, seem to have evolved into automatic egg-laying machines: they lay two eggs, incubate them for seventeen days, raise the squabs for two weeks, and then lay two more eggs – even though the young pigeons don't leave the nest for another week or ten days. Thus the immature pigeons help to incubate the new eggs, an arrangement that allows pigeons to continue reproducing in cold weather and accounts for their incredible productivity. Pigeons lay eggs every four or five weeks for at least nine months a year, sometimes even longer – there is a report from Winnipeg of pigeons incubating eggs on January 14, when the temperature was -14°C, and a friend of mine in Toronto who has an office on Bloor Street says the pigeons outside her window were laying eggs in December, when the temperature was -22, on top of the bodies of frozen hatchlings from a previous clutch.

This might argue against the intelligence of pigeons, but non-stop egg-laying is a common adaptation to superabundant food supply and relatively few predators – cats, Animal Control officers, TTC buses, and the occasional peregrine falcon (no kidding). Migrating birds fly south in the winter for the abundant food supply of the tropics, but fly north again to breed to avoid the numerous predators the tropical climate also affords. Birds that have abundant food supplies in the north eventually realize they don't have to migrate. To pigeons – and, increasingly, mourning doves – Toronto must seem an infinitely expandable niche.

In fact, pigeons are extremely intelligent birds. The psychologist B.F. Skinner chose pigeons to be the first subjects in his experimental Skinner Boxes – in which he developed his theories of behaviour modification and eventually raised his own children – because he believed pigeons to be more intelligent and adaptable than rats. Pigeons still replace guinea pigs in many psychological tests: in a recent experiment conducted at the University of Iowa, pigeons kept in Skinner Boxes were taught to recognize human facial expressions. They were shown photographs of human faces displaying fear, happiness, anger, and sadness and were rewarded with pinches of grain when they pecked the correct key. The same psychologist, Edward Wasserman, had earlier taught pigeons to organize pictures of objects in the same way humans would. The ability to recognize different expressions, says Wasserman, "requires very sophisticated nervous systems" and shows that human beings are not the only animals to communicate by such means. It remains to be proven that pigeons can actually *communicate* by means of facial expressions, but there are no doubt some pigeons in Iowa who think they've successfully trained human beings to supply them with pinches of grain on demand.

After the eggs are laid, both the male and the female share the task of incubating them – the male from mid-morning to late afternoon, and the female from then on overnight. In this way, the eggs are covered 99 percent of the time, and egg mortality is very low in both summer and winter. When the eggs hatch, the squabs are fed "pigeon milk" – a creamy excrescence regurgitated from the crops of both the male and female parent – for the first four days, after which the milk is mixed with semi-digested grain until, at day nine, the squabs get all grain until they leave the nest around day thirty-two.

Pigeons are also extremely susceptible to disease, the most common being a form of Newcastle disease called avian paramyxo. According to Paloma Plant at the Toronto Humane Society's Wildlife Department, where more pigeons are brought in than any other animal, the paramyxo virus is the number-one killer among urban pigeons. "We get about 400 cases of it a month," she says. "We call it the Linda Blair Syndrome, because its chief symptom is the head turned around and bent down, like Linda Blair in *The Exorcist*." A pigeon with paramyxo is always trying to see behind itself. Always. "When they get it they can't feed, they can't fly straight, they bump into buildings and break their wings." She shows me a pigeon that has just been brought into the centre's lab on River Street by someone who found it "acting funny," and sets it down on the floor. The pigeon immediately begins spinning around backwards, in tight circles, like a dog with fleas trying to bite its own tail. Paloma and her colleague, Wendy Hunter, look up at each other and nod.

"Paramyxo," says Wendy. "We have to euthanize it."

The pigeons in Sibelius Park – there are about fifty of them pecking at handfuls of seeds on a bare patch directly in front of the statue – look like the members of a symphony orchestra in grey tuxedos, all crawling around on the floor looking for their music. What is most remarkable about them is their incredible colour patterns: no two have exactly the same markings, a phenomenon common elsewhere only in mongrel dogs, human beings, and snowflakes. Darwin found the variation within domestic pigeon breeds – and the fact that traits of the original rock dove were repeated when different pure breeds were cross-bred – so absorbing that he based a large part of his theory of natural selection on a study of domestic pigeons.

More recent professional ponderers have used pigeons to study the evolutionary advantage of flocking: why do city pigeons, like the extinct passenger pigeons, fly in large flocks rather than singly or in pairs, as mourning doves do? It could be that fifty birds have a better chance of finding food than solitary birds do, but then fifty birds *need* to find a lot more food than individuals would, and mourning doves seem to have no trouble with food supplies. It is more likely that flocking has to do with defence against possible predators. In the wild, pigeons' main threat comes from the air, in the form of hawks, ospreys, and peregrine falcons. Once a prey species is numerous enough to be identified by a predator as a major food source, a given individual has a better chance of surviving an attack if it is part of a flock, especially if it is flying at or near the centre of the flock, than if it is flying along by itself. And studies of pigeon flock behaviour seem to bear this out. If you chart the flight pattern of each individual member of a flock of fifty pigeons, you will find that the birds are constantly changing position within the flock at regular intervals, so that each pigeon is at the periphery of the flock, where it is exposed to the most danger, the least possible amount of time. Thus all birds share the danger equally, but no one bird is exposed to attack for longer than is absolutely necessary.

I try to test this theory out with the pigeons milling about in front of Sibelius, to see whether each individual pigeon pecks its way in and out of the horde in any discernible pattern – to see, in other words, whether pigeons on the ground behave like pigeons in the air – but soon find my mind wandering. Fifty pigeons is not exactly a horde, I think. Come to think of it, 20,000 pigeons is a fairly small number for a city the size of Toronto. New York has an estimated seven million of them. Boston had 25,000 in 1966, when the city instituted its pigeon-poisoning program (corn soaked in arsenic) to protect the integrity of its historical architecture – a concern pigeon-proud London never seems to have had. Feeding the pigeons at St. Paul's Cathedral and in Trafalgar Square is as ancient and honourable a pastime as cavorting about the May Pole – as early as 1385, Robert de Braybrooke, Bishop of London, complained about "those who, instigated by a malignant spirit, are busy to injure more than to profit, and throw from a distance and hurl stones, arrows and various kinds of darts at the crows, pigeons and other kinds of birds building their nests and sitting on the walls and openings" of St. Paul's. London has had a by-law against molesting pigeons ever since.

Most other large cities now have some kind of pigeon program. Every day workers in Paris climb up to special dovecots constructed on the tops of the city's public buildings and collect pigeon eggs, a laborious but effective control: there are now about half the number of pigeons in Paris as there were just after World War II. Even smaller towns feel the need to keep their pigeon populations in check. In Simcoe, Ontario, a retired OPP officer named Bernie Goetz goes out every Sunday morning with a .22 rifle and shoots pigeons. The city pays him to do it: "I get a buck a bird," he says. "I usually try for a body shot. If you aim for their heads, they're just too small." He's been doing it since 1982 and figures he's killed about 3,000 pigeons. As Terry Looker, the chief inspector for the Ontario Humane Society, has commented, "We rather hoped people wouldn't do this." I wonder if anyone ever said that about passenger pigeons?

A PARLIAMENT
OF BATS

O ne night last fall, a bat crawled through a small hole in
the screen on one of our kitchen windows, no doubt
thinking it had found a safe, dark place for the night.
When we came down in the morning, it was curled up between
the screen and the glass; it couldn't have been more exposed if it
had been dangling from the clothesline. The small, dark-brown
and black parchment cocoon, clinging to the screen, looked
kind of sheepish. By the time we'd made coffee, it had begun to
edge its way back towards the hole it had crawled in through.
The hole was about the size of a quarter, and the rim was poked
inward, so that the bat would have had to pick its way over a
circle of needle-sharp wires in order to make its escape to the
outside. Watching it made my own stomach muscles flinch. It
gave up after a few half-hearted attempts and crawled back to
its corner, wrapped itself up in its leathery wings again, and
for the next three days pretended we didn't exist.

We weren't all that surprised to find a bat in our kitchen
window; we've known for a couple of years that our house has
bats. It's an old three-storey brick house, built in 1915 as near
as we can determine, and when we bought it there were a few
structural flaws where the roof joined the walls; the wood
around a south-facing dormer was rotted through, there were

gaps in the fascia boards, and a fist-sized hole in the brick-work just below the eaves where some earlier hydro wires had entered the house. All these, we soon realized, were bat high-ways. When some electricians we later fired hammered a three-foot hole through a wall at the top of the third-floor stairs in order to install a three-inch switch box, we had bats in the house every night for a week, zooming down the halls in a blind panic, circling around the bedrooms at ceiling level, getting caught in curtains. We took to keeping a fishing net beside the bed; I'd get up in the middle of the night, afraid to turn on a light because I couldn't remember if that would attract them or send them retreating to a dark crevice where I'd never find them, and I didn't really want to do either. I'd eventually see, or rather sense, a darker spot in the blackness swooping towards me and I would throw up the net in a kind of desperate self-defence gesture, but I soon learned that the best way to capture a bat is to wait until it settles on a door frame or window ledge, and just pluck it down like a piece of fruit. Then I would carefully march it outside and let it go, whereupon it would fly straight up to one of the gaps in our fascia boards and get back into the house.

We consulted an expert in bat removal. She came and looked through the hole the electricians had made and told us we definitely had bats in there. Then she looked at another hole in another part of the house and said we definitely had bats in there, too. Our house was more or less lined with bats. She said we probably had two separate colonies, with maybe 100 bats in each colony. She said she could poison them, but that would create quite a smell. She could scare them out with an electronic gizmo she had, but unless we sealed up their entry holes, they'd just come back. And if I sealed up their entry holes while they were still in there, they'd die and cre-ate quite a smell anyway. She more or less hinted that the best thing we could do about the bats was learn to live with them, which made me wonder how she made a living.

Eventually I patched all the holes between the inside of the house and the wall cavities. It was a kind of uneasy truce; we agreed to let them have the attic, the outside wall cavities and the floors in one or two rooms we didn't use very much, if they would agree to stay out of the main part of the house. So far it's worked; there hasn't been a bat inside the house for two years. We still hear them, especially during mild spells in the winter, crawling around in the third-floor walls and sometimes down into the living-room walls, where it's warmer. We've done some reading, and we know what they're doing in there, and we don't like it. But they're respecting the terms of the cease-fire, so we feel we ought to as well.

What we have are big brown bats (*Eptesicus fuscus*, which means "dark-coloured house-flyer"), one of eight species found in southern Ontario (there are 854 species worldwide). The other seven species are forest-dwelling bats, not usually found in the city in winter; the little brown bat (*Myotis lucifugus*), more common farther north than in Toronto, winters in caves and mine shafts in colonies of up to 300,000, so we should count our blessings. The red bat (*Lasiurus borealis*) a small, brick-red, fast-flying woodland bat, is migratory; it has been found in Toronto – the first one was recorded in 1854 by an entomologist who found it hanging from a tree in the Homewood Estate, north of Carlton Street, and who captured it and sent the skin to the celebrated naturalist Louis Agassiz – but only during rest stops in the spring and fall. Worldwide, according to York University's Brock Fenton, one of the most experienced and respected bat researchers in the world, only about twenty bat species are at all well known: "Ten percent of the genera are known from fewer than ten specimens, and quite a few species are known from only one or two animals." Unfortunately, big brown bats are among the species that have been well studied, otherwise we could donate our house to science.

One of the most remarkable books I've read about bats was written in 1952 by an amateur naturalist named Leonard Dubkin. Called *The White Lady*, the book is an account of his encounter with an albino little brown bat, born to a colony he had stumbled across in an abandoned field near the Chicago waterfront. Dubkin spent two years studying the colony, going down to their grotto under a locust tree, spending most evenings and often all night observing the roosting habits of the 600 bats belonging to the colony. The birth of the albino – apart from being an extremely rare and thrilling event in itself – allowed him to follow the growth and development of an individual member of the colony, something that would otherwise have been impossible without a radio collar or some other marking device that was unavailable to him at the time. Amateur birdwatchers have, over the years, contributed enormously to our accumulated knowledge of many bird species, but amateur batwatchers are almost as rare as albino bats. Dubkin found it unfortunate that so few people were interested in bats: "They are missing an aerial circus full of marvellous feats," he writes, "of daring stunts, and of hair-breadth escapes. Often a bat will dive into the air after a swift-flying bee. Sometimes a bat will flutter from side to side through the weeds behind a little blue butterfly, or dive straight at the trunk of a tree to pick off a resting fly. I have seen three or four of them dive at the same insect, and have held my breath in the certain belief that they were going to meet head-on in a rending crash. But somehow they never collide; they swerve suddenly at the crucial moment and go off in different directions."

The truth is that most people recoil at the thought of handling a bat. During the three or four days the big brown bat was in our kitchen window, various visitors and family members would look at it, shudder, and make a remark that usually included words like "creepy," "repulsive," and "disgusting." Bats are mammals; in fact, they are the second most numerous of mammals – at 854 species out of a total of 4,000 mammal

species, nearly one mammal species in five is a bat (the most
numerous, at 1,550 species, are the rodents). But bats are not
flying rodents, as they have sometimes been called, although
some of them may be distantly related to other insectivores
such as shrews and moles (some species of which can echolocate, but at vastly inferior levels than bats). At least one biologist has claimed that Australian fruit-eating bats, known as
flying foxes, are closer to primates than they are to other
mammals, a thought that has revived speculation about vampire bats. Accounts of the three species of true vampire bats
that are found only in Central and South America evoked horror among European readers of early travel and nature writing.
Captain Jared Stedman, for example, returning from a military
expedition to Guiana in the early nineteenth century, described waking up one morning to find his toes covered with
blood: "I had been bitten by the vampyre spectre of Guiana,"
he wrote, "which is also called the flying dog of New Spain;
and by the Spaniards, *perrovolador*. This is no other than a bat of
monstrous size, that sucks the blood from men and cattle,
while they are fast asleep, even sometimes till they die. . . ."
Stedman went on to say that these flying dogs made minute
punctures in the big toe and proceeded to suck blood through
the orifice until they were scarcely able to fly, "and the sufferer
has often been known to pass from time to eternity." The
noted nineteenth-century naturalist and explorer Henry Walter Bates, in his wonderful book *The Naturalist on the River Ama-
zon*, published in 1863, fed the flames of vampire phobia with
his account of being awakened one night while staying in "the
red-tiled mansion of Caripí," on the Pará River, "by the rushing
noise made by vast hosts of bats sweeping about the room":

> The air was alive with them; they had put out the lamp,
> and when I relighted it the place appeared blackened
> with the impish multitudes that were whirling round and
> round. After I had laid about well with a stick for a few

minutes they disappeared amongst the tiles, but when all was still again they returned, and once more extinguished the light. I took no further notice of them, and went to sleep. The next night several got into my hammock; I seized them as they were crawling over me, and dashed them against the wall. The next morning, I found a wound, evidently caused by a bat, on my hip.

Vampire bats do have a disturbing fondness for fresh blood, but they don't suck it from the carotid artery (or the hip) through hollow fangs, as Dracula seems to have done; rather they make a shallow gash in exposed flesh with their razor-sharp incisors (having first shaved away the fur or hair with their equally sharp cheek teeth), lap up the blood with their tongues, dog-like, when it comes seeping out, gorge themselves until they are so full of blood they can barely fly, then return to their roosts and regurgitate it for their young. The victim usually doesn't even wake up. The Red Cross could take lessons. But the Victorians can be forgiven for maligning vampire bats: even in their native ranges they had a somewhat sinister reputation. The Mayan god of the underworld, Zotz, is depicted as a human figure with the head of a vampire bat, holding in his hand a human heart dripping blood.

"Old myths die hard," as Brock Fenton notes in his book *Just Bats*. "For no better reason than that they are nocturnal and shy, bats are viewed with foreboding by a large segment of society; arguably, they are the most misunderstood, feared, and persecuted of mammals." Bats fly at night in search of insects, not blood (as many insects do), having evolved that way to minimize competition with insectivorous birds, but there is something disturbing to us about the thought of darkness filled with flying, furry things with claws and teeth; owls, one of the few carnivorous nocturnal birds (and a major predator of bats), have something of the same spooky reputation. Bats, by the way, do not want to become entangled in human hair, but when they do they are perfectly capable of extracting themselves from it; we know this because, in the eighteenth century, a British nobleman conducted an experiment in which he placed three bats in the hair of three of his female friends, and the bats removed themselves almost as quickly as the ladies did.

"That's one of the many misconceptions about bats," says Diane Devison, the senior biologist at Metro Zoo's African Pavilion and a long-time admirer of bats. "When most people think of bats, they think of blood, rabies, and hair, and there is really no basis for such fears. It's a matter of education." Diane became interested in bats when she was working in the zoo's nursery and found herself rearing a pair of tiny African fruit bats. "I found they were so smart, I could train them to fly to me for treats; I'd hold a piece of fruit and they'd fly over and sit on my finger and eat it. Eventually I took them home, and they'd fly around the house, come and perch on my shoulder while I was cooking dinner, or land on my desk when I was writing and try to pull the pen out of my hand."

When she delved more deeply into bat biology and evolution, she discovered, as Brock Fenton has also noted, that not a lot was known about bats except that they are intelligent and extremely well adapted to their various habitats (which, in evolutionary terms, is the same thing). Prejudice, she realized, begins with ignorance, so she began taking her fruit bats around to schools as part of the zoo's Outreach Program, showing kids that bats are not scary or dangerous or even particularly ugly. One day she found herself sharing a platform in Scarborough with a birder, and they both were talking about building special boxes to provide alternative housing for displaced birds and bats. "There were about forty kids in the class," she says, "and at the end of the talk we asked how many wanted to build bird houses and how many wanted to build bat boxes, and thirty-five of the forty said they wanted to make bat boxes. Great, I thought, this is headway. But two weeks later, when we all met again to put up the boxes in the Rouge Valley, thirty-five kids showed up with bird houses. I said, 'What happened? You all said you were going to make bat boxes.' And they said, 'Well, we wanted to make bat boxes, but our parents wouldn't let us.' So that's when I realized that it wasn't the kids I had to reach, it was the adults."

As a result, Diane became involved with such projects as *Cottage Life* magazine's 1993 trade show, in which she set up a booth to show people how to build bat boxes for their cottages – she had 950 Torontonians asking for bat-box plans in one day. Since then, the Metro Zoo (which sponsors bat conservation through its Bat Box Program) has had more than 4,000 requests for plans and information. "There is a lot of interest out there," Diane says.

One of the fears about bats does have some basis in fact: bats can carry rabies. According to Rick Rosatte, about 1 percent of insectivorous bats such as the big and little brown bats have the rabies virus, and since 1980 about 3 percent of humans treated for rabies have contracted the disease from bats. Of the twenty-five documented human deaths from rabies since 1925, says Rick, three died of bat rabies. But they weren't from Ontario – one occurred in Saskatchewan in 1970, another in Nova Scotia in 1977, and one in Alberta in 1985. Last year, a five-year-old girl in New York State died from rabies after she had been bitten by a rabid bat – she found it lying on a forest floor and picked it up and did not tell her parents when it bit her. "We tell people not to pick up sick-looking bats," says Diane.

Rick also points out that when a bat is diagnosed as rabid in Ontario, it's usually a big brown bat. "Even so," says Diane, "there is more danger of catching rabies from the family pet than from bats. Besides," she adds, "we don't encourage people to handle bats. We don't want people treating bats, or any wild animal, as pets. What we are telling people is that bats are here and they're here to stay, and we might as well learn to live with them. And building bat boxes is one way to make living with them fun and interesting."

Another myth about bats, says Diane, is that they are blind: "blind as a bat." Because they are nocturnal, their vision is better in poor light than it is in bright light, but they are far from blind. They see only black and white; their eyes do not have cones, or wide-spectrum refracting cells that allow us to see colours – but then, cones don't do us much good at night either. Nor do their eyes have a *tapetum lucidum*, the layer of reflective tissue between the retina and the eyeball that magnifies the amount of light that enters through the pupil – it is this layer that makes various animals' eyes shine at night when caught in the headlights of a car. We don't have them either. In fact, a study by John Pettigrew of the University of Queensland, Australia, found no fewer than thirty features related to bat vision that were thought to be unique to primates, which is why some researchers think fruit bats at least are descended from a long-extinct species of flying lemurs. Recent DNA studies tend to challenge Pettigrew's theory, but he seems to be hanging on to it anyway. "I'm not giving in just because a few people have sequenced a few genes," he told a newspaper in Dallas, Texas.

However they acquired their visual acumen, bats can see perfectly well after about twilight: experiments show that little brown bats can pick out objects two millimetres in diameter in very dim light. In bright light they can barely discern larger objects well enough to avoid flying into them; the bats careening down our hallway when I switched on the light must have felt a bit like I do when I'm swimming underwater without my glasses on.

Bats don't really need 20/20 vision; over the 20 million years since they first appear in the fossil record, they have evolved a better way of getting around in the dark. This peculiarity of bats was first noticed in 1790 by the Italian biologist Lazzaro Spallanzani, who spent the first half of his life proving that if you boiled swamp water for forty-five minutes, no micro-organisms grew in it. Turning his fertile mind to the question of night vision, he blinded several bats and found that they were able to fly about in the dark, and even catch their weight in insects, with undiminished efficiency. He then plugged the ears of the same bats with brass tubes and was surprised to find that when the tubes were open, the bats could fly normally, but when the tubes were closed off, the bats became disoriented and flew into objects placed in their path, after which he came to a perfectly logical, scientifically sound conclusion: bats could see with their ears.

Bats have, by and large, huge ears – one species found in Ontario, the northern long-eared bat (*Myotis septentrionalis*), has ears which, when bent forward, reach out farther than its nose, and the head of the spotted bat (*Euderma maculata*), a western species, looks like a pair of leather change-purses with a tiny face depending from them like a hasty afterthought. Hearing is obviously a highly developed sense in bats. But no one really knew what they were listening to until a series of experiments conducted in the 1930s by American mammalogist Donald R. Griffin, then at Harvard University. Griffin determined that bats issue a series of high-pitched squeaks, about ten per second during normal flight but up to 200 per second when zeroing in on moths or dragonflies, at frequencies up to 200 kilohertz, or ten times higher than human ears can detect. The echoes of these squeaks are picked up by the bats' ears and used to pinpoint the precise location and even shape of the object from which those echoes are rebounding. Griffin termed this ability "echolocation" – successive scientists sometimes use the term "biosonar" – and opened the door to one of the most fascinating aspects of study in the field of mammalogy.

The ability to echolocate at a bat's level of sophistication requires not only extra-sensitive hearing but also neural receptors in the brain capable of processing the information provided by the ears. Bats emit different kinds of squeaks – high-frequency, modulated-frequency, and a combined high-low squeak – each of which provides the brain with different information. The high-frequency echoes tell the bat how far away an object is; the combined-frequency squeaks allow the bat's brain to calculate the size of the object and perhaps its texture – which in turn would indicate whether the object was a moth or another bat; and the modulated frequencies create a kind of Doppler effect that indicates whether the object is moving, and if so in which direction and how fast.

All this information, computed virtually instantaneously in brains the size of a split pea, allows the bat to change flight paths and perform complicated manoeuvres with amazing agility. Although in a cramped hallway bats may seem awkward and blunderish, that's mostly because of panic created by bright lights and contradictory echoes; under normal circumstances, bats are more skilful flyers than birds are. Fixed feathers and the absence of teeth (teeth weigh a lot) give birds a slight advantage when it comes to speed and flight sustainability, but a bat's huge wing area, and its ability to manipulate individual sections of each wing give it amazing dexterity in the air; when you see a bat flying erratically, as poet James Reaney seems to have done, it's usually because it is chasing an insect:

He hangs from beam in winter upside down
But in the spring he right side up lets go
And flutters here and there zigzagly flown
Till up the chimney of the house quick-slow
He pendulum-spirals out in light low
Of sunset swinging out above the lawns.

The insects it chases range in size from grasshoppers and moths to caddisflies and mosquitoes – little brown bats eat up to 500 mosquitoes a day, big brown bats somewhat fewer, and Diane Devison says big brown bats in some Toronto houses may be switching to cockroaches, although they prefer insects that can be caught on the wing (Asian cockroaches, take note).

In southern regions, bats swoop above ant and termite colonies during the spring swarm, when millions of ant and termite nymphs are flying about looking for mates, but in Toronto, where termite swarming is less common, bats thrive mainly on flies and mosquitoes, which they catch by scooping them out of the air in their tails or wings, and then bending forward to eat them. Some bats cache their catch in cheek pouches for calmer delectation later, others devour them on the fly, like in-flight snak-paks. Since we can't hear bats when they echolocate, one way to check whether they are in your area is to watch a moth in flight during the late evening. Some moths have developed the ability to intercept a bat's sonar and will take evasive action when they do – such as suddenly folding their wings and plummeting to the ground. If you see a moth doing that, look around for a bat nearby. Tiger moths seem to have developed an even more effective defence against bats: they have a noise-making apparatus in their chests that emits sharp clicks that foul up a bat's detection equipment, giving the moth an extra split second to get away. In response to which, Canadian biologist James Fullard has suggested, bats may be developing a counter-strategy: emitting extra-high-frequency signals that are unlike those given off by tiger moths.

Big brown bats are the only species that hibernate in our houses and garages in Toronto. Little browns are hibernators, but perhaps because they are smaller and cannot tolerate extreme cold, they need deep, dark caves that never go below about 4°C and 100 percent humidity throughout the winter. Big browns seem able to withstand below-freezing temperatures and less humidity. Both manage to make it through to spring by reducing their heart rates from 200 beats per minute during normal, summer weather, to about 10 or 15 per minute in the dead of winter. Such a low heartbeat uses up almost no energy, which is essential because bats take in no energy while they are hibernating. (Which also means they do not defecate in their hibernacula, a bit of information we were quite happy to receive.) Ordinarily, bats lose about 25 percent of their body mass over the winter and emerge in the spring, when the ambient temperature reaches about 10°C, starving and dehydrated. According to Wendy Hunter at the Toronto Humane Society, long, cold winters can have a devastating effect on the big brown bat population. "They normally go into hibernation weighing twenty grams, and come out weighing about fifteen," says Wendy. "But after the cold, cold winter of '93-'94, for example, a lot of them came out at only ten or eleven grams. People were bringing them in here to the Wildlife Department left and right. The poor things were too weak to forage for insects; we had to hand-feed them on Espalac, a puppy milk replacement developed by dog breeders, to get their strength up."

Being that low on energy in the spring is hard on all bats, but especially hard on the females, who have to begin nursing young at this time. Bats mate in the fall (usually); the females and sexually immature males spend the summer days in segregated roosts, cool, dark hide-outs where they congregate to hang upside-down until feeding time in the evening. Sometime in September, the roosts are visited by the sexually mature males, and for a week or so the colony indulges in the kind of day-long, group activity that would have interested the author of the *Satyricon*. It certainly caught the attention of the author of *The White Lady*: during the second week of September, Dubkin noticed "a sort of nervous tension among the inhabitants" of the bat grotto. "Bats in pairs, always one behind the other, zipped nervously back and forth in the grotto, then suddenly hung themselves up from the roof." When he came back three days later, he watched the behaviour of one particular pair:

> The pair I had selected for observation hung side by side, so close together that their wings touched, like a boy and a girl holding hands. They licked each other with long red tongues – the head, the neck, the body and the sex organs, first one end and then the other. Then the male

let go his hold on the roof with one leg, twisted his body
until he faced the female, and put one wing around her,
exactly like a boy putting his arm around a girl to kiss her.
. . . Now the male let go his hold on the roof altogether,
twisted himself around onto the female's back and began
to copulate. Then the female did a strange thing. She put
her wings up to the roof, grasped the vegetation with her
thumbs, and pulled herself, with him still on her back, up
among the vegetation. Was it modesty that made her
want to hide herself while engaged in the sex act, or did
she prefer to copulate in some other position than hanging by her heels from the roof?

Modesty? Holy anthropomorphism, Batman. This sexual
behaviour in little brown bats has often been observed; Brock
Fenton has also watched little brown bats copulating later in
the season, after the colony has gone into hibernation. Every
now and then, he notes, a few male bats will wake up and
begin "moving from one cluster of torpid bats to another,
selecting individuals and attempting to copulate with them."
The accosted bats were often other males, he says, "and partners of either sex usually did not arouse from their winter
sleep during the encounter. In the active phase of mating,
males used distinctive copulation calls to calm struggling
females objecting to being mounted."
    Fenton goes on to lament the fact that almost nothing is
known about the mating systems of big brown bats, and I can
do little to enlighten him from our own experience. We can
attest that during warm spells in the dead of winter, at least
some of our big brown bats rouse themselves and crawl about,
up and down the walls, across the ceiling, accompanied by
much scratching and a great deal of disgruntled squealing.
This could mean that big brown males are copulating with
their drowsy male and female colony mates. I had rather
hoped they were chasing down pockets of cluster flies.

Once a female big brown bat has been impregnated, however and whenever that occurs, she does not become pregnant
immediately, but rather stores the male's sperm in her vagina
until spring. This ensures that mating can take place when
both the male and female are in top physical condition – after
a summer of good eating – and that the offspring is not born
during the winter months, when its chances of survival would
be nil. In May, the female's ovaries release exactly seven eggs,
which travel down to meet the stored sperm and become inseminated: by some mystical process, however, only two of
the seven inseminated eggs are allowed to progress towards
fetal development, and only one of them actually results in a
fetus that survives to term – a period of forty days. Like primates, female and male bats have only two nipples (male
Dayak fruit bats of Malaysia actually lactate – the only male
mammals known to do so – but male big brown bats do not);
the newborn bat immediately crawls up its mother's fur to her
breast, clamps its mouth over the nipple, holds on with its milk
teeth, and rides around with her in that position until ready
for independent flight in about eighteen to twenty-one days.

We decided long ago to keep our truce with the bats in our
house; maybe they are keeping our cluster-fly population down
in the winter, or maybe they are engaging in an endless, drowsy
orgy. We don't know; every now and then, when the noise
level gets too high, Merilyn pounds on the wall and shouts at
them, and they stay quiet for a while. They know the rules.
Diane Devison tells me about a large bat colony that was living in a few old pre-colonial houses in Chatauqua, New York.
When the houses' new owners renovated, the bats left and
never came back. "They just disappeared," she says. "No one
knows where they went. Did they panic? Did the colony split
up and take off in all directions? Or did they fly somewhere en
masse and establish themselves in a new community? Where
could they have gone?" Well, they could have come here.

Winter

Thou barren waste; unprofitable strand,
Where hemlocks brood on unproductive land,
Whose frozen air on one bleak winter's night
Can metamorphose dark brown hares to white!

– Standish O'Grady, "Winter in Lower Canada," 1841

URBAN
COYOTES

T his story could have taken place anywhere in Toronto, in any of the ravines, parks, cemeteries, or golf courses. It happened in February, so I could read it in the snow, but it might have been any time of year. A small grey rabbit was killed and eaten by two coyotes, and eaten so thoroughly that there was almost no trace of the event except for three sets of tracks, a disturbance of snow where the tracks converged, a few tufts of fur, and two drops of blood. No bones, no mess, as there would have been had the hunters been domestic dogs. Most people out for a morning walk would have passed the spot without noticing the story written there; I missed it myself until I saw the scat on the path about 200 metres farther on, recognized it as coyote, and retraced my steps to the beads of red and the bit of fur.

The rabbit's tracks met the path at right angles. It was probably an eastern cottontail (*Sylvilagus floridanus*), by far the most common lagomorph in southern Ontario, but there was too little of it left for a positive identification. It had been browsing in a patch of wild raspberries just off to one side, at the base of a steep hill leading up to a row of townhouses – I could see the tracks and the tell-tale raspberry stems nipped off at perfect 45-degree angles, the reddish stubble looking

213

like medieval spikes poking up through the snow. A few scattered brown pellets suggested the rabbit had spent perhaps fifteen minutes grazing on the sweet stems before hopping over to the path. The coyotes had been walking side by side along the path; I picked up their trails about thirty metres before the point where the rabbit stopped. There had been a light snowfall – the coyotes' tracks were set in a centimetre of downy powder, and filled in by a lighter dusting – and I could see how the tracks directly registered, the hind feet coming down on the imprint made by the larger forefeet, so that each track-line was a single, almost straight series of dots, right up to the point of intersection. A dog's tracks would not have looked like that; a dog's tracks would have been all over the path, zigging from side to side, zagging off the path to sniff at a deadfall, going all places and no place at once. Coyotes, when they walk, know where they're going and they go straight there. Their tracks did not deviate a centimetre until seconds before coming upon the rabbit. Perhaps the rabbit saw the pair and froze, hoping to dissolve into the morning light; or perhaps its back was to them as it cocked its large ears forward to listen to the traffic hissing by on the overpass above the ravine. In any event, there was no chase. The male coyote, travelling to the left of its mate, made the kill almost in passing, leaning over to its left to grab the rabbit by the back of the neck with its jaws, lifting it several centimetres off the ground and then snapping down hard, bracing itself with its right forepaw, which skidded a little on the path, while the female sat on her haunches and waited. She was two weeks pregnant. The pair were probably on their way home after a night of hunting meadow voles in Mount Pleasant Cemetery when the rabbit appeared as an unexpected treat. There was no desperation in the scene, no violence, no alarm. The coyotes ate the rabbit quickly and quietly, licking up the blood in the snow, crunching the small bones into one-centimetre pieces before swallowing them. The male probably ate first while the female kept watch, then she would take her turn.

Coyotes metabolize their food quickly, especially the indigestible bits, like hair and bone. The scat I found on the path – wild canines always defecate on paths, often where two paths cross or the path they're following forks, using their scat as a scent marker to tell other canines whose territory they are trespassing in – was almost two centimetres in diameter and about ten centimetres long, a small grey cylinder of grey rabbit fur wrapped around bits of bone. The fur protects the coyote's intestines from being damaged by the sharp bone fragments. I guessed it was the male's. The size of the scat and bone segments was at the upper limit for coyotes – a diameter of about two centimetres, or pieces of bone bigger than a centimetre, and I would have had to think wolf. These weren't wolves, they were eastern coyotes, Canis latrans, which means barking dog, although coyotes never bark in the city. Eastern coyotes are bigger than their western counterparts, and the males are bigger than the females; this male weighed maybe sixteen kilograms, the female fourteen.

Our word "coyote" is a sixteenth-century Spanish priest's attempt to render the Aztec glyph for the animal, which must have been pronounced coyotl, into Latin, something an educated Spaniard could read. The Aztecs incorporated the coyotl glyph into the names of several of their deities (including the one we call Montezuma, whom they called Motecuhzoma Xocoyotzin), indicating the high respect they had for a creature whose intelligence and survivorship must have reminded them of their own. The coyote cult moved north with the turquoise trade, turning up among the Anasazi of what is now New Mexico: the Anasazi were the ancestors of the Hopi and were pushed out of their pueblos in Chaco Canyon and elsewhere either by their own wastes – archaeologists have found huge garbage dumps in the canyons – or by the Navajo and Apache, northern Athapascan tribes that migrated south the same time the Anasazi disappeared. The Navajo, Apache, and Hopi all incorporated Coyote into their myths as the Trickster, the wily character whose job on Earth it is to make some sense out of the First One's creation. The Navajo word for coyote means "God's dog," the ancient palindrome.

Which is why I was intrigued when, a month after I found the rabbit kill, a coyote was hit by a car near the corner of Steeles and Leslie, and everyone who stopped to help it thought it was a dog. One of them, John Marquis – a student at the Ontario Bible College – said he thought it was "divine intervention" that made him pull over to help the animal drag itself off the road. No one expects to see a coyote in the city. We think of coyotes as desert animals. Or television characters. What was this one doing trying to cross a six-lane highway at rush hour? Someone named it Wylie when it was taken to the Toronto Wildlife Centre on Westmoreland Avenue. Its right hind leg had been broken, there was a deep gash on its left hind leg, and a lot of its hair had been scraped off – it looked like Wily E. Coyote on a bad day. Maybe the car that hit it was a Plymouth Roadrunner. At the Wildlife Centre, Leslie Cudmore treated Wylie for shock and transferred him to an animal shelter in North York, where technicians thought he would recover in seven or eight weeks. But ten days later his leg became infected, and they had to put him down.

Coyotes do venture into the city, and many of them now live here full-time. Dennis Voigt at the Ministry of Natural Resources office in Maple receives three or four calls a month about coyotes. Not long after Wylie was put down, Dennis heard about two more coyotes that were seen loping across the Mississauga Golf Course, and he was worried that they might drive out the three red fox families that lived there. "Foxes usually move out of territory claimed by coyotes," says Voigt. But a month later the foxes were still there and the coyotes had apparently moved on. They may have been members of the pack that hunts the Credit River system, or they may have been a pair of transients come down from the area between the MNR offices in Maple and the 401, a wide swath of farmland in which coyote sightings are a regular occurrence. "We know there are den sites in the Don Valley," Voigt says. "We've had reports of coyotes as far south as the Leslie Street Spit. There is plenty of good coyote habitat in Toronto. What we don't know is whether the coyotes in the valley are permanent city residents, or whether they drift down into the city from north of the 401."

Coyotes ordinarily form packs of about five or six animals, made up of a dominant or alpha pair, a non-mating beta pair, and one or two pups that are either too young to leave the pack or, for reasons of their own, have decided to stay on after the normal time of dispersal, which is November. In the wild, such a pack will patrol from fifteen to fifty square kilometres of home range, which they demarcate from the home ranges of neighbouring packs by scent marking – urinating on trees, rocks, branches, tufts of grass, cached food sites, or defecating on trails. Each pack has its own distinctive scent, and some researchers think each coyote has his or her own scent; whatever the case, there is no doubt that a coyote scent post is a complex melange of pheromones that relate a detailed story to other coyotes. Some scent posts say there is food or no food along this path. Others warn that the pack's denning or core area is nearby and will be vigorously defended. Transient coyotes picking up such messages immediately turn tail and leave, not wanting to enter into spirited territorial debates. The transient will often be escorted to the pack's perimeter by a sentinel coyote.

The idea of sentinel coyotes is not fanciful. Colleen Campbell, a volunteer coyote researcher I met in Banff, Alberta, told me that one day in May she and park warden Mike Gibeau went looking for the den of a radio-collared female they'd been keeping an eye on throughout the winter.

The female had been killed in the town of Banff, run down by a car — a common fate of urban coyotes — and when Mike examined her body he discovered she'd been lactating. He and Colleen had gone to her core area to find her den, because they knew there must be five or six pups waiting for Mama to return. "We just wanted to see if the rest of the pack was looking after them," Colleen said. "We were walking along a dried-up stream bed, following a nice set of tracks, when suddenly we heard a coyote yapping at us up ahead. We looked up and there was a young male, sitting beside the trail, sort of waiting for us to notice him. He stood up and moved a few yards farther down the trail, then turned around to see if we were following him. Then he sat down and yapped again, and moved a little farther off. I swear he was trying to lead us away from the area, because the pups were there. At least we knew that the pack was taking care of the pups, so we just turned around and left." Later that summer, after the surviving pups would have vacated the den, Colleen and I went back and found the den, about nine metres off the trail, exactly where the sentinel coyote had been sitting when she and Mike had visited in May.

In cities, packs tend to be much smaller, often just the mated pair. Like all wild canines, coyotes mate for life (we have bred that trait out of our domestic pets, just as we are breeding it out of ourselves). The female chooses her mate very carefully from a suite of admirers. This selection process takes place in mid-December, when the female will be followed about for weeks by half a dozen males, each of which tries to demonstrate his suitability by bringing her food or growling menacingly at his rivals. Gradually, one or two of the males lose interest and drift away, and near the end of January the female indicates her choice from among those that remain. She does this by approaching the male, lying down on the ground, wriggling playfully under his chin, reaching up to lick his muzzle, and conducting an assortment of other unmistakable signs of affection. Sometimes the male will respond with a long, extended, ecstatic howl, and sometimes the female will join in to make it a duet.

Mating takes place in early February. The female is in estrus for only one week of the year, so readiness is all. In a pack of six or seven, only the alpha female goes into estrus; the beta females may engage in sexual activity, but they do not become pregnant unless the population is threatened by food shortages or predation – if a local sheep farmer, for example, is out trying to exterminate the pack. The gestation period is sixty-three days – as it is for most canines – so the female dens up in early April, and the pups are born two or three weeks later. This means that the pups are born when the winter is becoming milder and food sources more plentiful.

By the time the pups are old enough to leave the den, in June, the weather is fine, the food supply – mainly rodents and lagomorphs and, in some areas, white-tailed deer fawns – is in full swing, and conditions are optimal for the pups' survival. Even so, from a litter of six or seven pups usually only two make it through the summer. Coyotes have few natural enemies – owls and hawks will occasionally swoop down on a frolicking pup at night, lynxes and bobcats will steal into the den and make off with a pup or two, but in the wild the chief cause of pup mortality is starvation, and the viruses to which semi-starvation makes them more susceptible: rabies is rare among coyotes, but canine hepatitis is common. In populated areas, according to Mike Gibeau, the main killer (after cars) is canine distemper transmitted by domestic dogs. Mike conducted a three-year study of urban coyotes in Banff, in order to determine how many coyotes were making the townsite part of their regular beat. He put radio collars on eleven coyotes; within a year, seven of them had been killed by motorists. But he did find out that coyotes don't come into town any more than they have to. "If part of the town occupies 10 percent of their normal territory," Mike says, "then the coyote spends 10 percent of its time in town and the rest in the other 90 percent of its range. They aren't coming into town and staying there to feed on garbage, or cats."

The coyote den Colleen and I found had been dug in sandy soil, on a south-facing slope under a fallen tree and near water – in fact it was on an island in the Bow River. A pack's core area will contain several den sites – as many as ten – and the female will move her pups from one to another for no obvious reason – maybe just to be on the safe side, or maybe to get away from den parasites such as fleas and ticks. The openings to the dens are neither concealed nor obvious; small holes about thirty centimetres in diameter, often defined by spaces between tree roots or fallen logs that look for all the world, when you walk by them, like natural cavities in the loose soil until you look more closely and realize that one of them takes a sharp turn and continues deeper into the hillside – another one I examined in Yellowstone National Park went in about eight metres. Dens usually have one or two other exits that are better concealed, often by grassy hummocks or some other vegetative cover. The Don Valley is rife with dozens of excellent den sites, but so far none have been seen in use. Which is not to say they are not in use, only that coyotes that do not want to be seen are not seen.

When the pups are born, the female remains in the den with them continually for the first week, and the alpha male and beta coyotes bring her food: whole rabbits or rodents, or partly digested meat from larger kills, such as deer, which they swallow and regurgitate in the den. When the pups are weaned, they share in the regurgitated meal: I have come across coyote scat in the spring that consisted entirely of an intact field mouse – the coyote must have swallowed it whole and then waited a little too long before returning to the den to feed the pups. At three weeks, the pups begin to leave the den, although they return to it to sleep or, if their mother gives the alarm signal – a quick, imperative yip – to hide from danger. They are weaned at six weeks, and the adults gradually increase the size and complexity of the prey they bring to the den – from dead field mice to live rabbits – so that the pups learn not only how to eat solid food, but also how to catch it. Pups begin hunting grasshoppers; they stalk and pounce on them in fearsome imitation of the way they will stalk and pounce on field mice as adults.

Dens are occupied only during pup rearing. When pups leave the den for good in July, the coyote pack resorts to its usual habit of sleeping in lays – open hollows in the ground, or depressions in tall grass that they create by turning around in tight circles until they have a bed – the origin of the domestic dog's dim practice of spinning around on the sofa before lying down on it. The pack remains together throughout the summer, mostly resting and playing during the day and hunting at night. They hunt small rodents almost incessantly; studies have found that a coyote will make an attempt at field mice or meadow voles about once every twenty minutes, with about a 33-percent success rate, which means a single coyote will eat fifteen small rodents, or half a kilogram of meat, per day. The senses they use during hunting are, in order of importance, hearing, sight, and smell – the sight of an alert coyote pouncing on a grass-covered vole tunnel is one of the most thrilling in nature.

Coyotes are not desert animals, particularly; their ancestral homelands are the Great Plains, from about Mexico up to just north of the Canadian border. Charred coyote bones have been dug up in native campsites near Saskatoon dating from about 1500 B.C. Like us, they are an animal of the arid grasslands, a locale they shared with bison, jackrabbits, and pocket gophers. Since they couldn't bring down bison, they quite sensibly developed a taste for jackrabbits and pocket gophers. When Europeans eliminated the grey wolf from the hills and forests surrounding the Great Plains (a process that coincided with Europeans' intention to raise domestic livestock on the

Great Plains, despite the Great Plains' outrageous unsuitability for domestic livestock), the coyote moved into the niche vacated by the wolf – except that, by and large, it continued to prefer rodents and lagomorphs to larger game. By then, it had evolved a propensity for solitary hunting rather than hunting in packs, which makes it much more able to cope with changing food sources than the wolf. Wolves hunt in packs because one wolf cannot pull down a moose. Coyotes hunt singly because it doesn't require a pack of them to pull down a Richardson's ground squirrel.

But coyotes are cunning and infinitely adaptable, and in the hills and forests there did appear the occasional large carcass – wolf-killed deer, aged moose, spindly elk calves – upon which opportunistic coyotes were happy to feed. Although coyotes on the Prairies consume almost nothing but rabbits and rodents, about 90 percent of the food eaten by coyotes in Riding Mountain National Park in Manitoba consists of moose carrion left behind by wolves. For coyotes, nothing is carved in stone. As we eliminated wolves, cougars, and grizzlies from most of their former range, the wily coyote moved in; and as we cut down trees and planted grass – creating ideal coyote habitat, because rodents love grasslands as much as coyotes love rodents – coyotes moved in again. Our efforts to stop them have not succeeded: there are more coyotes now than ever before in their history. As Ernest Thompson Seton wrote in *Lives of the Hunted* (1901), coyotes have persisted in spite of our efforts to stop them: "They have learned the deadly secrets of traps and poisons, they know how to baffle the gunner and hound, they have matched their wits with the hunter's wits. They have learned how to prosper in a land of man-made plenty."

By 1875, they had expanded from the Midwest into the Rockies, north into Alaska, and east to Minnesota. They entered Ontario in the 1890s, travelling north through Michigan into southwestern Ontario, then into eastern Ontario in the 1920s, and Quebec and the eastern United States by the

1940s. A separate wave seems to have come from Manitoba, along the north shore of Lake Superior, and down into southern Ontario. All the coyotes now found in eastern North America, then, came via Ontario. Wherever we created coyote habitat, coyotes moved into it. Farms, parks, country estates, the grassy verges of highways, suburban lawns: coyotes view them all as fairly faithful miniature reproductions of their native Great Plains.

During their sojourn in Ontario, coyotes seem to have picked up some wolf genes. Ontario has two distinct types or subspecies of grey wolves (*Canis lupus*): the boreal wolf is found in the northern part of the province, north of Lake Superior; and the Algonquin wolf, a slightly smaller version, so-called because it has been studied mainly in Algonquin Park, is found in southwestern and eastern Ontario as well. Coyotes and grey wolves can interbreed (so can coyotes and domestic dogs, but not as successfully), and in areas where wolf populations were dwindling and appropriate mates becoming more and more difficult to find, wolves turned to coyotes for mates. The Algonquin wolf seemed pre-adapted for interbreeding with coyotes: it is only slightly larger than a coyote, had already adjusted to life in the farmlands of southern Ontario, and hunted small prey such as rabbits, groundhogs, and rodents (although its preferred food was still deer – Algonquin wolves probably came down into southern Ontario following the deer herds). Some researchers think that red wolves in Texas are also a cross between grey wolves and coyotes, and some sheep ranchers hope they are, because cross-breeds are not covered by the Environmental Protection Act.

Although coyotes have been well established in rural Ontario since the turn of the century, their appearance in the city is a fairly recent development. At first glance, a city isn't perfect coyote habitat. All of a coyote's principal enemies are in it, for one thing: human beings, their cars, and their pets' diseases. The ravines and parks offer some den sites – in Banff,

I saw a female coyote raising her pups in a culvert that ran under a four-lane highway – but not enough to lure coyotes out of their natural areas, and few large enough to enable coyotes to pack up and defend fifteen-square-kilometre home ranges. Toronto's slightly warmer climate may be a partial draw – concrete and asphalt form "heat islands" that are on average two or three degrees warmer than their surrounding hinterlands, a temperature differential significant enough to make life tolerable to many species of insects, algae, and viruses, but not to many mammals. And yet, coyotes are moving into cities, including Toronto, in unprecedented numbers. So what's the attraction?

The answer is plentiful food sources. I don't think, as many pet owners seem to, that coyotes are coming into Toronto specifically to eat cats and small dogs. As a rule, predators do not hunt down other predators for food. Just look at us – our prey species are almost all herbivores. True, we do eat fish, which are carnivorous, and dog is a popular dish in Asia – I still recall seeing rows of skinned, barbecued puppies hanging from wires in market stalls in Beijing. North American Natives ate dog meat during religious ceremonies – coincidentally, their dogs were probably semi-domesticated coyotes. But by and large, we tend to eat ungulates and kill carnivores only to eliminate them as rival predators (sometimes we call it sport, but it has the same effect). There is no reason to assume other motives in coyotes: when domestic cats are out at night, they are hunting, usually for mice, chipmunks, and other small rodents – the same prey sought by coyotes – and when a coyote finds another predator trying to occupy its niche, trouble ensues.

Coyotes are becoming more common in cities as cities spread out into what used to be the country. Residents of suburban areas talk about coyotes "moving in" to their parks and golf courses, but those parks and golf courses used to be pastures and clearings, parts of coyotes' rodent-hunting ranges. In many parts of suburban Toronto, the coyotes have been there all along; it is the human encroachers who have moved in.

Toronto is a veritable coyote smorgasbord: house mice and voles in the residential areas, white-footed mice in the parks; chipmunks and groundhogs in the ravines; ducks and Canada geese in the wetlands; and grey squirrels everywhere. Domestic cats probably take care of most of the smaller rodents, although our tendency to declaw our pets in homage to our furniture may be giving coyotes an advantage. But the larger prey species, the groundhogs, the shorebirds, the Norway rats and the squirrels, are pretty much left to coyotes and red foxes. According to Dennis Voigt, there are about 1,000 red foxes living in Toronto; I doubt there are that many coyotes – although there are easily enough grey squirrels to support such a population. But 100 coyotes in the 800 square kilometres that comprise the Toronto conurbation – fifty mated pairs – would have 16 square kilometres each in which to forage for game, about the same area as they are used to in the wild.

Also, Toronto's coyotes don't have to depend on rabbits and rodents to survive in the city, because coyotes will and do eat anything. Studies of suburban coyotes in California in the 1980s found that garbage made up 78 percent of the diet of coyotes foraging in the Los Angeles suburb of Glendale, and only 2.5 percent for coyotes in Claremont, another suburb, where a local by-law required homeowners to stow their garbage in spring-lidded metal containers. Claremont coyotes gorged on rabbits, wood rats, birds, grasshoppers, and avocados. Elsewhere, coyotes have been found with bellies full of wheat, pups cut their teeth on grasshoppers and crickets, and adults at certain times of the year feed almost exclusively on blueberries and acorns. They are opportunistic feeders, and as William Jordan, author of *Divorce Among the Gulls*, points out, intelligent, opportunistic animals are the kind that are most likely to succeed in the big city: "Most

urban wildlife," he writes, "are characterized by general habits
and catholic tastes. A generalist is adaptable. In nature it lives
in a range of environments. It is usually an omnivore, eating
everything from vegetables, grass and nuts to freshly killed
insects, birds and mammals, to carrion of various stripe." Jordan might have been looking at a coyote as he wrote that.

People tend to become nervous at the thought of a pack of
wild coyotes running loose in the city, like a motorcycle gang
crashing a lawn party. Parents of small children imagine their
youngsters beset by ravenous animals. This concern springs
from two sorts of misinformation: one, that coyotes are really
wolves in disguise; and two, that wolves (and therefore coyotes) attack human beings without provocation. Jack London
and Walt Disney are responsible for the second myth: legends of Lobo and White Fang notwithstanding, the number
of documented cases of unprovoked attacks by wolves on
human beings over the past century can be counted on the
fingers of one hand. (Never mind for now the number of
cases of unprovoked attacks by human beings on wolves.)
And the number of coyote attacks on human beings is proportionately smaller (there are a few more cases, but there are
a lot more coyotes). About a dozen people a year are beset by
coyotes in Southern California, and recently a pair of coyotes
attacked a family in Vermont. In 1985, two teenagers were
bitten by a coyote in Banff – one was dragged out of her tent
by her hair and suffered minor lacerations to her face and
forearms. The coyote was tracked by park wardens and shot.

Such cases are serious, but consider the statistics for
domestic animals. In Southern California, where twelve people are bitten by coyotes, 26,000 are bitten by dogs every
year, and no one launches campaigns to rid the country of
Dobermann pinschers and pit bull terriers. According to
William Winkler of the United States Centres for Disease
Control in Atlanta, each year in the city of New York 40,000
people are injured by their family pets seriously enough to

require medical attention. The Humane Society reports that in the United States and Canada, more than two million cases of attacks on human beings by domestic animals are reported every year. In the two years covered by Winkler's study, eleven people were killed by their own pets. In those same years, not a single case of attack by wolves, coyotes, or bears was reported to anyone. I don't think we should be worried about coyotes; I think we should keep an eye on Spot.

At any rate, coyotes are not wannabe wolves, even though in rural Ontario they are still referred to as "brush wolves" and often simply as "wolves." I suspect calling them wolves makes hunting them more exciting. There is, however, no compelling reason to transfer our psychologically suspect fear and loathing of wolves (imported largely from Europe anyway) onto coyotes, unless one happens to be a sheep farmer. Fortunately for coyotes (and for sheep), there are very few sheep farmers in Toronto.

Having coyotes around can be a good thing, even for farmers. Coyotes eat mice, and mice eat grain. In the early 1940s, the citizens of Klamath County, Oregon, learned the hard way that eliminating the weight on one side of nature's balance can have unforeseen repercussions. Over a period of several years, they killed more than 10,000 coyotes that they thought had been preying on their sheep and cattle herds. By 1947, the coyote had been extirpated from the county; but in the absence of coyotes, the field mouse population went crazy, shooting up to 25,000 mice per hectare. Mice ravaged the grain crops, and the cost in lost grain revenue far exceeded the previous cost in lost sheep. The coyote was reintroduced. More recently, coyote eradication programs instituted in the southwestern United States, where 90,000 coyotes are killed by government hunters every year, resulted in an explosion of deer mice in 1990. The deer mice were found to be carriers of the rare hantavirus, which causes severe respiratory disease in humans: dozens of people died before the virus was even identified, and hantavirus has now spread north. Two people have died from it in British Columbia, one in Alberta, and there have been several cases reported in 1994 in Ontario. Perhaps this is not a good time to be eliminating coyotes.

Besides, I rather like the idea of coyotes cruising through Toronto's green spaces, sniffing at gopher holes, driving German shepherds mad by scent-marking their telephone poles. Coyotes, inhabiting a place in our psyches between wolves and domestic dogs, are a true link with wild nature. Although they do not howl or bark in the city, I have heard coyotes howling at night in the wild, and felt the kind of visceral kinship with them that human beings have felt for canines since we first left the trees and ventured out into the grasslands. I picture a mated pair of coyotes bedding down at the same ravine off the Don Valley where Ernest Thompson Seton built a makeshift cabin when he was ten years old, a spot not very far from where I found the scant remains of a coyote-killed rabbit, and which he calls Glenyan in his novel *Two Little Savages*. Seton may have seen coyotes in the Don Valley — in those days, the valley was virtually wild north of Dundas, and in his largely autobiographical novel he describes a hunter in what is now the Rosedale Ravine shooting a lynx. "To a small boy as I was then," he wrote in 1938, "it was a far cry into a wild and distant country." He certainly later encountered western coyotes and felt the same subliminal affiliation I did: "If ever the day should come," he wrote, "when one may camp in the West and hear not a note of the coyote's joyous stirring evening song, I hope that I shall long before have passed away, gone over the Great Divide." Seton went over the Great Divide in 1946, but I'm sure he would have rejoiced at the thought of coyotes loping through the same woods he did as a child and felt the kind of astonishment any generous host feels when a guest, though uninvited, turns out to be an interesting and beneficial friend.

STURNOPHOBIA AND

THE BIRD OF AVON

A t 7:30 in the morning of August 24, 1920, James Henry Fleming, of 267 Rusholme Road, Toronto, was out in his back garden when he saw a flock of seven black-coloured birds flying overhead, just clearing a row of tall elms on the west side of the road and disappearing off towards Ossington. He was so excited by this sight that he hurried into the house and made a meticulous note of it in his diary. "I was at the east end of the garden," he wrote, "when the birds were first seen directly above me, and I was able to watch them for nearly three hundred feet of their flight. . . . The shape of the birds, their flight, and the movements of the flock were characteristic," he continued, "and I had no doubt, while the birds were in sight, of their identity." They were, he confidently stated, European starlings.

Starlings. Why would anyone make such a scrupulous fuss over a flock of starlings? Most of us see hundreds of starlings every day. Starlings are one of the most common birds in Toronto, even in August. The mystery deepens when we consider that the writer, James H. Fleming, was at the time one of the most notable ornithologists in North America. He was the son of a commercial seed-grower whose one-hectare plot occupied most of the corner of Yonge and Elm streets, and he had begun collecting in his father's seed garden in 1884, at the age of twelve (his first trophy was the eggs and nest of a vesper sparrow). By the time of his death in 1940 he had amassed 32,267 items – stuffed birds, bird skins, nests, and eggs – the largest private collection in the world. When Fleming bequeathed it to the Royal Ontario Museum upon his death, ROM became the possessor of one of the most important ornithological collections in North America. His "Birds of Toronto" had appeared in the bird-science magazine *Auk* in 1906 and 1907, he was named Honorary Curator of Ornithology of the National Museum of Canada in 1913, and from 1932 to 1936 he was president of the American Ornithological Union. In short, Harry Fleming was Toronto's, and perhaps Canada's, best known bird man. So why was he so excited at seeing a handful of starlings flying over his garden in August 1920?

Because it was the first time anyone had seen starlings in Toronto. Like house sparrows, European starlings (*Sturnus vulgaris*) are not native to North America but were introduced on this continent by well-intentioned but ill-informed amateurs who thought it was a good idea at the time. An organization known as the Society for the Acclimatization of Foreign Birds, headquartered in New York, believed that if there was any bird in the world worth having, the United States should have it. Its model was the Acclimatization Society of Great Britain, formed in 1860 and dedicated to the importation of non-British animals into England, largely for commercial purposes, the idea being that tapirs, for example, could be raised for profit by British farmers as easily as pigs. A public dinner sponsored by the society in 1863 featured such exotic menu items as kangaroo steamer, Syrian ham, curassow (a wild pea-hen from Paraguay), Honduras turkey, leporin (a Central American fish, also called headstanders), and Canada goose, but at its height it never had more than a hundred active members. The society died of silliness in 1866, while contemplating the introduction of Japanese deer and Catalonian asses – and leopards and jackals to keep the populations from exploding. But before it died, it spawned offspring clubs in the colonies. The Acclimatization Society of Victoria, in Australia, for example, brought the roe deer, the partridge, the rook, the English sparrow, and – its most celebrated import – the rabbit to that benighted continent. In North America, Society members liberated a handful of European starlings in New York's Central Park in 1877, but the release was not successful – the birds refused to nest in their foreign home and died without producing a second generation. Another liberation was attempted in Portland, Oregon, two years later and was no more successful than the first.

But the spirit of acclimatization was on the land. During the 1880s, a literary-minded New York drug manufacturer named Eugene Schieffelin, who had founded the American Acclimatization Society, decided it would be nice if every species of bird mentioned by Shakespeare existed in North America. Accordingly, he imported and released, successively but not successfully, flocks of chaffinches and song thrushes (*Winter's Tale*), nightingales (*Romeo and Juliet*), and sky larks ("Sonnet 29"). In neither case did he import enough birds to form a viable breeding population. He was luckier, however, with the European starling (*Henry IV Part I*), sixty of which he released in Central Park in the summer of 1890, followed by forty more the next year. The starlings were immediately successful, and it was noted with some satisfaction that the first starling nest in North America was found in the central tower of the building occupied by the American Museum of Natural History. It seemed a good omen.

The starling quickly expanded its range and, curiously, altered its habits. It became a predominantly urban bird, and it ceased migrating. In Europe, starlings had been the particular study subjects for biologists studying migratory behaviour in birds – several banding projects in Germany, for example, had tagged thousands of German starlings and charted their migration routes into Africa. But in North America, the transplanted starlings seemed not to know where to go, so they went nowhere. They did, however, multiply. In 1924, the brilliant amateur naturalist Margaret Morse Nice, spending a year with her husband in Amherst, Massachusetts, noted that in January "there were few birds but Starlings," and that since the birds had shown up in that city in 1910, they had ceased to migrate. Nice wasn't very interested in starlings – she was then becoming known as a specialist in song sparrows – but there were no song sparrows in Amherst in January, so she noted starlings.

Nice's half-hearted view of starlings may have had something to do with their reputation. Despite their lofty literary beginnings, the starling was almost immediately branded a pest by ornithologists because of its tendency to oust native songbirds from their nest cavities. Eastern bluebirds, for example, were hard hit. They had already had to contend with the introduction of the house sparrow; now they had to fend off the larger, stronger, and more aggressive starlings, and birders feared this new aggressor could be the last straw. Few birds are as belligerent as the starling in taking over territory – they've even been known to eject woodpeckers and flickers from holes those chisel-heads had just finished making for themselves. Several prominent birders – Mabel Wright and Frank Chapman, founding editor of *Bird-Lore*, among them – lobbied the government for control programs, but the starling's insect diet made it a beneficial bird to farmers, and the Department of Agriculture recommended that starlings be protected by law. Birders knew better. Fleming may have been excited by his sight of starlings in Toronto in 1920, but two years later, when Percy A. Taverner published his monumental *Birds of Eastern Canada*, he looked upon the bird as a gardener might look upon crabgrass: "As though acclimatization societies and others with similar aims had not done enough harm already in introducing the House Sparrow to America," Taverner lamented, "they have added another factor of unknown possibilities in the form of this bird, to compete with native species, develop unforeseen qualities in its new surroundings and in the absence of its natural control to increase enormously." Taverner obviously expected the worst. He didn't refer to Fleming's sighting – in those days, a bird had actually to be "taken" to be counted as a resident – but noted that "after a few preliminary sight records in southern Ontario, it has finally been taken at Kingston." This momentous event took place on Wolfe Island in 1921, when Harley C. White of the Department of Mines captured a lone starling on the island: White sent the skin to Taverner for the National Museum of Canada's collection. The next year, Kingston ornithologist Edwin Beaupré saw three starlings in the city; in 1923 he saw four more. "Very soon after this," Helen Quilliam records in *History of the Birds of Kingston, Ontario*, "on March 27, 1923, he saw two small flocks, and by December 5, 1924, a flock of 150 near the city," and the bird was officially recognized as a breeding species. Meanwhile, starlings began moving into other parts of southern Ontario, where they were received with equal lack of enthusiasm. "It becomes my sad duty," E.M.S. Dale wrote in the *Canadian Field Naturalist* in 1923, "to chronicle the arrival of the European Starling at London, Ontario." Clearly, the starling was moving in, and moving in fast. Beaupré collected seven starling eggs from Wolfe Island in 1926 and sent them to the Royal Ontario Museum as proof that the starling had achieved breeding status in Ontario. As Taverner had noted sourly in 1922, "any hope of its eradication at present probably is vain."

By 1933, Torontonians were complaining about the proliferation of starlings in Lawrence Park, where 5,000 birds were noisily disturbing residents, messing up benches, and desecrating picnic spots. City workers were brought in to evict the squatters with shotguns, firecrackers, and even fire hoses.

Bombs were set off in trees, hourly, all night, terrifying babies and war veterans and blasting limbs off trees, until finally the residents sensibly decided that the sweet murmurations of starlings were preferable to the constant din of heavy artillery. Besides, the bombardments weren't working: the starlings would start into the air, circle patiently for a few minutes, and then either settle back down on the mutilated trees or else drift off to another part of the park, or to another park altogether. But the war continued into the 1950s: Des Kennedy, in his book *Living Things We Love to Hate*, recalls growing up in suburban Toronto and regularly watching "a couple of dozen men lined up along the street wielding shotguns, a policeman at their head. At a sign from the cop, the men would raise their guns, aim into the dense canopy of plane trees overhead, and blast away in unison. Huge thunderheads of starlings would erupt from the trees, screaming in protest." Still, Torontonians have not had as much reason to complain about starlings as Bostonians have; in 1960, a flock of 20,000 starlings got caught in the intake of a Boeing Electra jet engine at the Boston Airport; the plane crashed, and sixty-two passengers were killed. Or Vancouver, which in 1982 estimated its starling population at four million, most of which had chosen the girders of the iron bridge directly above the trendy shops on Granville Island as a roosting site.

Starlings and house sparrows are year-round residents in Toronto. On a sunny day in mid-winter, if you see a dark-coloured bird sitting on a chimney, or the top of a tree, looking somewhat like a crow only smaller, chances are it's a starling. Starlings are largish (twenty centimetres) blackbirds with short tails, long, dark-coloured beaks (in winter) and white speckles spread in a regular pattern on their necks, backs, and chests. The speckles are the tips of certain feathers that have turned white, and they have given the bird its name, since someone once decided that they looked like little stars ("starlings") set in an inky sky. Their black feathers are iridescent to

the point of looking oily, with bluish or copperish hues flash-
ing throughout their upper bodies as the sun hits them. Towards spring, their beaks begin to turn bright yellow and most
of the white speckles wear off. In northern regions, they are
most commonly confused with grackles, which are much bigger, do not look like oil slicks in the sun, do not have white
spots in winter (and in any case are rare in Toronto in winter),
and never have yellow beaks. Starlings belong to the family
Sturnidae, from *sturnus*, the Latin name for the starling; grackles are Icterids, a name that is derived from the Greek word for
jaundice, and are thus related to the orioles, so named because
of their yellow plumage.

Starlings roost in any place that causes inconvenience to
people. The branches of trees under which people have to
walk, sit, or park their cars are favourite places. Tremendous
flocks of them will choose a roosting site for no other apparent reason than that there is a child's sand-box or swing set
under it. There they will set up their murmurations to the
intense dismay of every human within earshot (or, sometimes,
buckshot).

Nest sites are chosen by the males and are generally holes
in trees or houses, five to eight metres off the ground, with
openings about 6.5 cm in diameter and enough room inside for
a nest with floor area about 150 square centimetres, or the size
of a paperback. The male builds a rudimentary nest in the cavity before he has a mate, then perches outside and flutters the
tips of his wings, puffs out his chest feathers, wipes his beak
on the branch, and sings his heart out until a female shows interest, either in his performance or in the suitability of his nest
site. Once a female has chosen him, she promptly throws out
most of the junk he has stuffed into the cavity and rebuilds the
nest with her own material, finishing it off with a soft interior
lining of feathers or plant down. The male apparently does
not take this as a personal affront and probably did a lousy job
on the nest on purpose, expecting his work to be tossed out.

The female lays a clutch of four to six eggs (six being the optimum number) in April or May, then moves to another nest in June and lays a second clutch. The hatchlings emerge on the twelfth day and three weeks later are ready to leave the nest, so by the time the female moves to her second nest, the young from the first have fledged. The female does most of the incubating – the male sits on the eggs 30 percent of the time during daylight hours, and never at night. Both sexes develop incubation patches, featherless areas on their chests that are particularly well vascularized to keep the eggs warm. After the chicks hatch, the parents spend almost all their time caring for them – making 300 food trips a day (200 by the male, 100 by the female) to keep the chicks supplied with insects (which in April is no easy feat in Toronto). With adults living an average of fifteen to seventeen years and raising ten to twelve young per year, it doesn't take long for starling populations to go geometric.

Which explains why there are now an estimated 200 million starlings in North America. The Christmas Bird Census for 1974, the first year pigeons were included in the count, found that there were more than three times as many starlings in Toronto as there were pigeons and house sparrows, and six times as many starlings as chickadees (and only twenty grackles).

The starling's success is also due to its intelligence and adaptability. The former is signalled by its uncanny ability to mimic other birds vocally – fifty-six other species, at last count – as well as a variety of other sounds. Its nearest relative is, after all, the mockingbird. I once listened to a starling perched on a neighbour's chimney going through its repertoire and distinctly heard it imitate a wolf-whistle, a cat, and a Bavarian cuckoo clock. I've also heard a starling down by Harbourfront make a noise like a gull. Others have reported hearing them sound like dogs, wind chimes, flutes, and a man with catarrh clearing his throat. Mozart is said to have had a caged starling that could whistle the aria "Trostlos schluchzet Philomele" from his opera *Zaide*, and may even have inspired it, even though the aria is supposed to be sung to a caged nightingale. Shakespeare's reference in *Henry IV Part I* to a starling (the Bird of Avon) is to one that will be "taught to say nothing but 'Mortimer'" – a sad under-employment of its prodigious talents – and Lawrence Sterne, in *A Sentimental Journey* (1768), has a caged starling save his hero from a trip to the Bastille by chirping "I can't get out – I can't get out," at a dramatically significant juncture.

A wide variety of songs is usually considered an advantage during courtship, since studies show that the most successful males are those with the biggest repertoire – although why the sound of a cat or a cuckoo clock is attractive to female starlings is anyone's guess. Vocal mimicry, however, is a subject that has been widely studied and hotly debated. Its purpose may be to divert predators as well as to attract females; the tendency of the starling to add a few trills and arabesques of its own to the songs of the mimicked species suggests that its intention must not be to fool a member of the mimicked species. In any case, starlings devote a large portion of their brains to memory storage. A comparative study of western and eastern marsh wrens showed that western marsh wrens, which are polygamous, have vocal repertoires that are three times more extensive than those of eastern marsh wrens, which are monogamous, and devote 50 percent more of their brain's storage space to vocalization. Eastern marsh wrens concentrate their mental power on food gathering and chick-rearing. Starlings, which are generally monogamous, seem to have the mental capacity to embrace both wide vocal repertoires and efficient domesticity, thereby making them twice as likely to pass their genetic material on to future generations. Which is probably why there are more starlings around than marsh wrens.

As the references to starlings in Shakespeare and Sterne suggest, keeping starlings in cages was at one time a common practice. The birds were captured when they were young and impressionable – there is an old English proverb that maintains, in effect, that you can't teach an old starling new tricks. Konrad Lorenz, the well-known authority on animal behaviour, had "an extraordinarily understanding friend" in England who described the starling as "the poor man's dog." Starlings and dogs, according to Lorenz, have much in common, in that both must be obtained as young animals if a true human-to-animal bond is to be established. "It is seldom that a dog, bought as an adult, becomes really your dog," he writes in *King Solomon's Ring*. Similarly, starlings must be caged as nestlings, and "you must feed and clean your nestling yourself, if you want a really affectionate bird of this species." This may require a lot of work, he says, but not for long. "A young starling needs for its development, from its hatching till it is independent, only about twenty-four days. If you take it at the age of about two weeks from the nest, it is early enough and the whole rearing process takes a bare fortnight. It is not too troublesome and demands no more than that you should, with the aid of a forceps, cram food, five or six times daily, into the greedily gaping yellow throat of the nestling, and, with the same instrument, remove the droppings from the other end."

As I say, if a starling can survive that, it can adapt to anything.

And starlings do, for the most part, adapt to new situations quickly. One of the most pressing problems faced by songbirds these days, perhaps even more pressing than the disappearance of their natural breeding habitat, is parasitism by the brown-headed cowbird. Cowbirds do not build their own nests, but like cuckoos lay their single egg in any convenient nest they can find, and the adoptive parent birds usually wear themselves out trying to feed the large, voracious cowbird hatchlings as well as their own gaping horde of young. Even without cowbird chicks to feed, 300 trips a day to feed their usual brood of six chicks takes an incredible toll on the starling's energy supply. Adding a larger, more aggressive cowbird chick to a brood of six starlings would obviously be disastrous for the parents and some, if not all, of the starling chicks. Starlings, therefore, have adapted by laying fewer than their optimal six eggs. A study conducted in New Jersey investigated 138 starling clutches: 73 of them contained only five eggs; 33 contained fewer than five, which means that 80 percent of the clutches examined contained fewer than the optimal number of eggs. Obviously, starlings have figured out that, brood parasitism by cowbirds being a fact of life (in some areas it happens 33 percent of the time), laying fewer of their own eggs to compensate for it would result in more of their chicks surviving. It seems to have worked.

One of the reasons cowbirds like starling nests is that both cowbirds and starlings are prodigious eaters of insects. An examination of the stomach contents of 2,157 starlings has shown that 57 percent of their diet consists of insects, predominantly beetles (including the marauding Japanese beetle), crickets, grasshoppers, millipedes, and butterflies and cutworm moths, compared to only about 8 percent for house sparrows. This amounts to nearly twenty-three grams of insects consumed daily, or nearly one-third an adult male's body weight. Much of the remaining forty-three grams, unfortunately, is taken up by berries and cherries, a factor overlooked by the U.S. Department of Agriculture when it gave starlings a thumb's up for aiding farmers. Starlings are not beloved by orchardists. Starlings especially like sour cherries and berries, as they cannot digest sucrose. They share this trait with a few other birds, such as robins and house wrens: like all vertebrates, birds are able to absorb sugar through their intestinal walls and into their bloodstreams by means of special enzyme-like mechanisms called transporters, which break down the complex sugar sucrose into more readily digestible glucose and fructose. Starlings have fewer transporters than most birds and so tend to avoid fruit with a high sucrose content. This is probably because such a large proportion of their diet consists of insects, and sugar transporters aren't vital for their survival: lipids from insects are stored more efficiently for long-term use than are sugars from fruits. Still, starlings are considered major pests by most fruit-growers. In response, re- searchers are looking into the possibility of biologically engi- neering blueberries to increase their sugar content, a measure that would make them less palatable to starlings and even more palatable to humans than they already are.

This seems a much preferable control measure to that envisioned by Timothy Findley in his novel *Headhunter*. Star- lings have, in general, not fared as well in recent literature as they did in Shakespeare's day. In Findley's novel, which takes place in a futuristic Toronto that bears a striking resemblance to Orwell's *1984*, flocks of starlings are deemed carriers of the dreaded disease sturnusemia, a bacteriological plague that is ravaging the city; squads of face-masked Metro employees cordon off trees containing roosting starlings, spray them with a lethal chemical, and then sweep up and burn the corpses. There was a time when some Toronto residents, especially those living around Lawrence Park, would have cheered such drastic measures, but so far – fortunately for starlings – cooler heads have prevailed.

RATS

WITH GOOD PR

L ast winter, as I was walking along the path leading to
our cabin, I watched one of our resident squirrels do a
very peculiar thing. The snow along the path was fresh
and light, about thirty centimetres deep. The squirrel was
perched at the top of a maple beside the path, about twelve
metres above the ground. I'd been watching it as I walked,
wondering what it was up to. It scurried back and forth along
the tree's bare top branches, stopped to look down, scurried
some more, and then, when I was about six metres from the
base of the tree, it suddenly launched itself into the air, arms
spread-eagled like a free-falling parachutist. It half-sailed,
half-plummeted to the ground, landing in the snow beside the
path with a muffled whoompf. Then it picked itself up and
made a dash for a nearby pine, leaving an impression in the
snow like that of a cartoon character crashing through a wall.
Even its tail had left a mark, a thin, curved line fringed by the
tiny striations of its guard hairs. I stopped in complete amazement as my brain tried to figure out what I had just seen. No, it
hadn't been a flying squirrel, it had been a black squirrel. No,
it hadn't been aiming for a neighbouring tree and missed;
there weren't any neighbouring trees close enough. Why had-
n't it simply climbed down the maple and walked across the

path to the pine? Apparently, it preferred to fly, for the sheer, unnecessary joy of it: Hey, watch this! I continued on my way feeling that I had been granted a rare glimpse into a secret life. It felt like a privilege.

"Tree rats," my friend Jamie Swift calls them. "Rodents with good PR." Another friend, David Macfarlane, wrote recently in *This Magazine* that "the sky-rocketing squirrel population of metropolitan Toronto has confounded biologists, puzzled statisticians, infuriated gardeners and raised the spectre of intolerance in the city's once-peaceful backyards and vegetable plots." In David's article, vigilantes prowl Toronto's suburban neighbourhoods with flame throwers, customs officials confiscate mail-order crates of exploding chestnuts. "The best way to deal with squirrels," says one of his characters, the bibulous Colonel Charles Bewlington, raising his squirrel gun, "is to aim just in front of them."

    Another friend, novelist Daniel Poliquin, has a theory about the difference between black and grey squirrels that is not flattering to either. His theory, elucidated in two of his novels, most recently in *The Black Squirrel*, is that black squirrels are actually crosses between grey squirrels and Norway rats. At some point in the not-too-distant past, rats began to worry about urban rat-extermination programs, and noticing how easy grey squirrels had it (being called cute and fed peanuts by children and senior citizens in parks), the rats struck a bargain with the grey squirrels: the two species would cross-breed and become black squirrels. "The crossbreeding had worked," he writes. "Ever since then, citizens and tourists fed them peanuts instead of cyanide pellets. But they were, in fact, just former rats that had preserved nothing but their colour and traces of a foreign accent." Daniel doesn't explain what the squirrels got out of the bargain, or why they went along with this plan. Maybe the rats talked them into it. Or maybe they wanted to become more aggressive, less twitchy, more rat-like. If that's the case, they got a bum deal, because recent studies show that black squirrels are just as twitchy as grey squirrels when chased by dogs or researchers blowing police whistles. In any case, Daniel's novel is really about the gradual anglicization of Ontario's francophones, so the whole story has to be taken metaphorically.

Except by Jamie Swift. "That would explain it," he says. "They really are rats in trees. I was right."

Partly right. Squirrels are rodents, members of the Sciuridae branch of the Rodentia family tree. Other sciurids include the groundhog, the prairie dog, several ground squirrels, chipmunks, and the flying squirrels. Grey squirrels are one of the six arboreal members of the family; they live in trees rather than in burrows, like the Richardson's ground squirrel and the eastern chipmunk. They much prefer to travel by leaping from tree to tree like the Flying Wallendas than by slinking along the ground. So they are, in a sense, tree rats, or at least tree rodents.

They certainly behave like rodents and are treated as such by farmers. Peter Kalm, the Swedish biologist who travelled throughout eastern North America in the 1750s, wrote that grey squirrels, which he called *Sciurus cinereus* (after Linnaeus, his professor back in Sweden), "frequently do a great deal of mischief in the plantations, but particularly destroy the corn, for they climb up the stalks, cut the ears in pieces and eat only the loose and sweet kernels inside. They sometimes come by hundreds upon a cornfield and then destroy the whole crop of a farmer in one night." As a result, various control methods were instituted. "In Maryland," he related, "everyone is obliged annually to kill four squirrels, and their heads are given to a local officer to prevent deceit. In other provinces, everybody who kills squirrels receives twopence apiece for them from the public on delivering the heads. The skins are sold, but are not much esteemed," he adds, noting elsewhere that squirrel skins

were used to make women's shoes. "Squirrels are the chief food
of the rattlesnake and other snakes, and it is a common fancy
with the people hereabouts that when the rattlesnake lies on
the ground and fixes its eyes upon a squirrel, the latter will be
as if charmed, and that though it be on the uppermost
branches of a tree it will come down by degrees till it leaps
into the snake's mouth." This sounds more like the Serpent in
the Garden, but Kalm *was* travelling through Puritan New
England. "The snake then licks the little animal several times
and makes it wet all over with his spittle," continues this student of Linnaeus, "so that it may go down the throat easier. It
then swallows the whole squirrel at once. When the snake has
made such a good meal it lies down to rest without any concern." This observation is supported by Mrs. John Graves
Simcoe, who recorded in her diary for June 29, 1793, that in
Niagara she "saw a stuffed rattlesnake, which was killed near
Queenstown in the act of swallowing a black squirrel." This
would have been an eastern massasauga rattlesnake, still
found along the Niagara River, but not in Toronto.

Kalm also noted that squirrel meat was considered "a
dainty" in New England, and Major Samuel Strickland, who
came to Canada as a colonist in 1826, confirms that the practice of eating black squirrel meat was still common then. "The
flesh is excellent eating, far superior to that of the rabbit. In a
good nut-season, in the western part of the province, the
quantity of these animals is incredible." And some things
never change: "These pretty little creatures," he writes, "are
very destructive among Indian-corn crops. I have seen them
carrying off a whole cob of corn at once, which I will be
bound to say was quite as heavy as themselves."

This business of eating squirrels is far from an historical
curiosity. Ernest Thompson Seton included a chapter on grey
squirrels in volume four of his *Lives of the Game Animals*, published in 1929, and a friend of mine who is a chef in a well-
known and extremely refined restaurant tells me that when

she was growing up in Ohio, she regularly ate squirrel meat. Her father would shoot them in the fall, she said, and in years when nuts were plentiful, their flesh was rich and nut-flavoured, "like that of free-range chickens raised on walnuts." (Shooting squirrels in Ohio is a traditional sport. In 1862, when the state was threatened with invasion by Confederate troops, 50,000 Ohio men volunteered for a regiment that became known as the Squirrel Hunters. Perhaps David Mac-farlane's Colonel Bewlington remembers them.) And there are still restaurants in some parts of the southern United States in which "Brunswick Stew" is a specialty menu item. Brunswick Stew is properly made with squirrel meat, and although some reprinted recipe collections, such as *The Williamsburg Cookbook*, contain the arch footnote, "Chicken is now substituted for squirrel," don't you believe it. Two years ago, the *Economist* reported that the cook at the Coon Club, a restaurant in Uniontown, Kentucky, regularly tosses young grey squirrel briskets into his soups and stews, or serves them deep-fried in flour and butter, with mashed potatoes and gravy. The meat is said to have "a sweetish taste." Uniontown is on the banks of the Ohio River.

Both Kalm and Strickland agreed with Linnaeus that the black squirrel and the grey squirrel were two separate species, *Sciurus niger* and *S. cinereus*, respectively, but it is now known that the two are in fact one species, the black squirrel being but a black or melanistic morph of the grey, which is *S. caroli-nensis*. The grey squirrel actually has several colour phases, depending on geographical location: there are grey grey squirrels, dark-brown grey squirrels, reddish-brown grey squirrels, and, near the town of Olney, Illinois, a large popula-tion of albino grey squirrels. When I lived in Chaffey's Lock, Ontario, north of Kingston, people would come from great distances to see Chaffey's grey squirrels, a significant number of which are black with white tails. Yet another friend, the poet Robyn Sarah, who lives in Montreal, tells me that she has never seen a black squirrel in that city; they are all grey except for a small albino population that lives on the top of Mount Royal.

Most of the grey squirrels in Ohio are grey – except for a black colony on the campus of Kent State University, the descendants, I'm told, of a pair brought there by a Canadian professor in the 1970s. If so, then the professor was probably from Toronto. Most of Toronto's grey squirrels are black. According to Eric Gustafson, an ecologist at White Pines College in New Hampshire, "the black morph was predominant in some areas prior to the advance of civilization," but has "declined throughout the northern half of the range of the grey squirrel since the late 1700s." This does seem to imply that Toronto has been relatively untouched by the advance of civilization, but the curious fact is that black squirrels still predominate in most urban centres, especially in the northern extent of their range. Gustafson and his colleague, Larry Van-Druff, found that black morphs made up 60 percent of all the grey squirrels in Syracuse, New York, although they made up only 2 percent in the outlying rural areas. No one seems to know why this is, but speculation on the subject has led to some fairly arcane, not to say squirrelly, research.

It is known, for example, that melanistic traits in red foxes are often accompanied by alterations in the function of the animals' thyroid and adrenal glands, as well as in their pituitary weight, which can cause a decrease in the fox's fright response. In other words, red foxes with some black genes are calmer. They frighten less easily. This evidently is not an evolutionary advantage in red foxes, more a genetic curiosity, otherwise there would be a disproportionate number of black red foxes around, and there aren't. But in grey squirrels, being slightly less twitchy may have its advantages. When I think of the black squirrel I saw sailing blithely off the top of a twelve-metre maple to do a belly-flop in a shallow pile of snow – the equivalent of me leaping off my cabin roof onto a Journey's End pillow – I think that is not the behaviour of an overly cautious animal. But is it an evolutionary advantage?

Well, maybe. To appreciate how, we need to know something about the grey squirrel's life cycle, especially as it pertains to reproductive behaviour. Fortunately, this is an interesting subject.

Grey squirrels are sociable animals. They each have their own well-defined home ranges, but these ranges overlap considerably, and one grey squirrel seems to tolerate another recognized grey squirrel coming into its territory without doing much about it. This is unlike red squirrels, for example. A red squirrel (*Tamiasciurus hudsonicus*) screeches like mad whenever another red squirrel dares to look crossways at it; grey squirrels quite happily share their foraging territory with others of their kind. Red squirrels live exclusively in coniferous forests – there used to be a lot of them in Toronto, when there were a lot of pines, but there are none now – and differ from grey squirrels in several other ways as well. Red squirrels tend to store all their food in one giant cache, for example. They'll spend the summer and fall dropping hundreds of pine cones down a single hollow tree, or stuffing them in a cavity under a stump, whereas grey squirrels, which live mainly in mature hardwood forests, bury nuts and seeds separately, one to a hole. No wonder red squirrels want to keep other red squirrels away: if a larger, more dominant red squirrel finds a cache and claims it, its original owner will starve to death over the winter. That isn't likely to happen to a grey squirrel: as the owners of Barings Bank were reminded recently, it's best to spread your assets around.

Even though their territories overlap at the edges, grey squirrels don't like to be overcrowded. In the wild, their average density is about one squirrel per hectare. In the city, that tends to increase a bit. Okay, it increases a lot. When grey squirrels were introduced into Stanley Park in Vancouver, they quickly settled in at about twenty per hectare. Toronto's population isn't that dense; studies conducted in the late 1970s in Mount Pleasant Cemetery came up with 127 squirrels in an area of about twenty-eight hectares, or about 4.5 per hectare. The area also had 7.4 largish trees per hectare, so that breaks down to about one squirrel for every two trees.

Grey squirrels are polyestrus – they can have more than one litter of pups per year. In fact, each female has one litter per year, with half the females in a colony mating at the end of January, and the other half at the beginning of June. Their mating ritual follows a four-stage pattern that has only, to my knowledge, been described once before, by D.C. Thompson of the University of Toronto's Forestry Faculty – the biologist who conducted the Mount Pleasant studies in 1974 and 1975 and published his findings in 1977. Thompson divided the squirrels' reproductive cycle into four distinct phases: the prechase, the mating chase, copulation, and post-coital behaviour.

The prechase, or "sexual trailing," is initiated by males probably in response to some preliminary release of pheromones by a female. The male approaches the female, sniffing tentatively, and the female moves away. The male approaches again, and the female moves away again. After a few approaches, she may allow the male to sniff the base of her tail, or she may not. There is no aggressive behaviour on either part – the male does not insist, and the female does not snap or snarl. This begins about five days before the female goes into estrus – in Thompson's study, it began on January 25 and May 21 – and becomes more frequent as estrus approaches. The chances of the male's getting close enough for a good sniff increase as the days go by.

The mating chase begins about seven o'clock in the morning on the day of estrus. It starts with one male chasing one female, but other males are attracted by the action, and the whole thing soon escalates into a mad parade of males, a whole hierarchical convoy, streaming through the trees behind a not-overly panicky female. Squirrels can run up to twenty-four kilometres an hour, and frequently do. First the female gives a kind of involuntary quack, a sound that tells the nearest male that she is in heat. He responds with a sort of whick, or what has been described as "a stifled sneeze," and the chase is on. Before long, there are five or six males involved, some attracted from as far away as half a kilometre. The first male is not necessarily the one at the head of the pack; in fact, he hardly ever is. In all of Thompson's observations of mating chases, it was the males who came from farthest away who were the most successful. And these were also the oldest males in the queue – no male younger than fifteen months was ever successful; the average age for the ultimately successful male was twenty-seven months.

The chase goes on for a brief time, with a few melodramatic incidents to break up the repetition. The female might dash into a tree cavity, for example, and the males will all line up outside waiting for her to come out, jostling one another for position as they wait. No one dares to go in after her. Then she'll dash out past them all, and the chase resumes, this time perhaps with a different lead male. Eventually, the female will put on a burst of speed and outdistance her pursuers. As soon as she is out of sight, however, she stops running, assumes the mating position, and waits for the lead male to catch up to her. When he does, they copulate and the other males sort of drift away as if they had something better to do anyway. The curious thing here is that the female leads the chase around her own home range and always stops to wait for the lead male almost exactly at the centre of her territory. You'd think that after a few chases one of the subordinate males would catch on and simply go to the centre of the female's territory and wait for her to arrive. But no, there is honour among squirrels, and adherence to tradition. Besides, the whole point of the exercise is to ensure that the strongest, fastest male – not necessarily the cleverest – and one from the greatest distance away, and therefore least likely to be a near relative of the female, wins the chase and gets to pass his genes on to the next generation.

As for post-coital behaviour, there isn't much of it. The female frees herself from the male's grasp and runs up a tree. The male calmly cleans his genitals and then follows her, making sure no subordinate males are lurking about with malicious intent, and, in at least 30 percent of observed cases, copulates with the female again after an abbreviated, some-what pro-forma chase.

All the males taking part in the chase are "recognized" males, that is, they all belong to a large and somewhat loosely defined colony in an area of about a square kilometre. A rec-ognized male can wander pretty much anywhere in that area without being harassed by other squirrels. If he wanders into an adjoining area, where he is not recognized, he will not be welcomed. He will not be allowed to take part in the chase. He may even be attacked and made the object of a chase scene himself. Female grey squirrels do not like intruders. Intruders make them jittery, and jittery females do not repro-duce well. This is true of many animals – sheep farmers will tell you that the mere suspicion of a coyote in an area will cause a ewe to lose her lamb. It might almost be said that the purpose of recognizing other squirrels and forming a colony, even a loose one, is to keep the females from getting jittery and thus to increase the likelihood of successful reproduction.

But in a city, where squirrel density is at least five times greater than in the wild, it must be difficult to maintain a colony's integrity. Unrecognized males must be wandering in and out of other squirrels' territories all the time. Females must exist in a more or less permanent state of the jitters. Now, sup-pose melanistic grey squirrels are genetically inclined, be-cause of alterations in their adrenal, thyroid and pituitary production, to be less jittery than grey grey squirrels. Suppose black grey squirrel females do not skip an estrus cycle every time they see an unrecognized male squirrel in their vicinity. That would give black grey squirrels a decided reproductive advantage over grey grey squirrels and might just account for the predominance of black morphs in Toronto.

A grey squirrel's gestation period is forty-four days, so given peak mating dates in Toronto of January 31 and June 5, most of Toronto's squirrels are born on March 16 and July 19, give or take a day or two for leap years and bank holidays. Grey squirrels like to live in hollow trees and tree cavities in the winter, because such places hold the warmth better, but in summer they make nests. You can see these nests better in the winter, when the trees have lost their foliage; they look like huge balls of dead leaves, usually stuck near the very tops of deciduous trees. When I was younger I thought these were crows' nests, but my daughter corrected me some years ago. She sometimes counts them when we go for walks and usually gives up at around twenty because there seems to be no end of them. Despite their haphazard appearance, they are very well made, which is why they last through the winter. First a foundation of interwoven twigs in the fork of a branch, then nipped leaves with their stems interlaced, enough to form a large ball with a hole in it at branch level. Inside, the nest is lined with chewed vegetable matter. The whole thing is soft and warm and water-tight.

In this nest, or in a similarly lined tree cavity in winter, the female gives birth to three pups, each weighing about sixteen grams. The pups are blind and helpless at birth and remain that way for nearly five weeks, when their eyes open. During this time the female nurses them and carries them wrapped around her neck when she goes out to forage for food. You can see them with a pair of binoculars and some patience if you sit in a park in early May or September. After seven weeks the pups are weaned, and shortly after that they are out foraging for themselves. This is a very convenient disposition of things, because it means the young of the year are foraging at the end of May and the end of September, when the maple samaras (we used to call them helicopters) are ripening, in the case of spring pups, and most nuts and acorns are falling off the trees in the case of the fall litter.

About 65 percent of these young of the year die in their first year, which is why we are not, contrary to the opinion of some of my friends, knee-deep in squirrels. Many of them fall out of their nests. One April day, when I dropped into the Toronto Humane Society's Wildlife Department, Wendy Hunter was walking around with a tiny, five-week-old squirrel in her hand. She was feeding it with a stopper full of Espalac, holding it up against her body in a face cloth to keep it warm. Its eyes were still glued shut, but it was sucking away content-edly. "Later," she said, "I'll soak peanut-butter sandwich cubes in milk and give it some of that, and then eventually just the peanut butter, then some fruit and veggies. Finally I'll give them Rodent Chow, which is mostly sunflower seeds. If I don't get it right, they'll have soft bones, which will break easily. Squirrels are the most common mammals we get in here," she added. "Especially this time of year. They climb out of their nests, fall to the ground, and kids come along and find them and don't know what to do with them, so they bring them to us. The best thing is just to leave them alone – the mother will come down and find them." Wendy makes sure she finds out exactly where the young squirrels are picked up, right down to the exact tree if possible, so they can be released back into their natal units. But obviously most of these bite-sized pro-tein packets are eaten by cats, raccoons or gulls before they can be rescued by the people who bring them in to Wendy.

Youth mortality isn't the only control on squirrel numbers. About half of all the adults in a given squirrel population die each year, either from exposure (mange, or the loss of hair caused by an infestation of mange mites, is a bigger problem in the city than it is in the country), starvation during low mast years, or being run down by cars as they try to cross a street from which the trees have been removed. Quite a few are lost to predation, mostly by domestic cats and dogs, some by hawks. One fall, when I was crossing the Bloor Street Viaduct on my bicycle, I saw a beautiful red-tailed hawk sitting calmly on the bridge's parapet, its back to the traffic, gazing north along the Don Valley. Another time, on Leslie Street Spit, an American kestrel glided across my path with something black and bright red dangling from its talons that looked a lot like part of a young squirrel.

Lack of food probably isn't as frequent a problem in the city as it is in the wild. During the eighteenth and nineteenth centuries, grey squirrels would periodically migrate in vast numbers from one part of their range to another. Awestruck travellers wrote of seeing thousands of grey squirrels swarming over trees and fields, and even fording rivers, in search of food. W.N. Blane, for example, a British merchant who undertook *An Excursion Through the United States and Canada During the Years 1822-1832*, wrote that he could scarcely believe his eyes when he "saw the immense number of these animals. . . . I found that this host of squirrels had in many places destroyed the whole [corn] crop, and that the little fellows were sometimes seen, three or four upon a stalk, fighting for the ear. One party of hunters, in the course of a week, killed upwards of 19,000. In most places, however, there were such multitudes of them that the inhabitants quite despaired of being able to rid themselves of this plague." Ernest Thompson Seton, in volume four of his *Lives of the Game Animals*, writes that some of these migrations originated in Ontario: "Innumerable Squirrels swam across the Niagara River," he reports, "and landed near Buffalo, New York, in such a state that the boys caught them in their hands, or knocked them from the fences and bushes with poles."

These mass migrations – which ought properly to be called emigrations, because they were not annual, and the squirrels did not return to their places of origin – were caused by crop failures. A late spring frost, a dry summer, an infestation of insects, any one or a combination of factors could mean a poor fall harvest of tree seeds and nuts, and when that happened, the squirrel population took off en masse in search of greener pastures (or corn fields). Such invasions no longer take place, and most researchers attribute this to human alteration of the deciduous forest. Until about the late 1700s, it was said that a squirrel could travel from the Atlantic coast inland to the Mississippi River without ever having to touch the ground; today a squirrel is lucky to be able to cross a street. We have, in the process of turning our mixed hardwood forests into shopping malls, inadvertently placed severe constraints on the squirrel population: they no longer swell to such numbers that regular crashes occur. Squirrel numbers, despite all appearances, are fairly stable, fluctuating by about 25 percent from year to year. A study conducted in a Delaware woodlot found that in high mast years the squirrels in the fourteen-hectare study area numbered 116; in low mast years they dropped to 82.

During high mast years, when nuts, acorns, and seeds are plentiful, grey squirrels seem to prefer hickory nuts above all others. Squirrels are omnivores, technically speaking. They eat nuts and acorns in the fall, but the rest of the year they'll eat buds, berries, pine seeds, bark, even grasshoppers, caterpillars, and birds' eggs. Park goers in Toronto may want to add peanuts to this list, but studies show that peanuts do not contain enough nutrients to support a grey squirrel. Then again, one imagines, neither does pine bark. Peanuts have a high caloric count, good for cold weather: the bird seed we put out in winter has a lot of peanuts in it, and it's always the peanuts that go first. But of all available foodstuffs, squirrels prefer hickory nuts. In a forest of mixed hardwoods, with oak and maple and hickory, squirrels will strip the hickory trees first and only then will they switch to the oaks. There are two kinds of hickories common in Toronto, the shagbark hickory (*Carya ovata*) and bitternut hickory (*C. cordiformis*), both of which produce nuts with hard shells, and so squirrels need to expend more energy to get at the meat than they do for acorns, but they seem to think it's worth it. And hickory nuts do have more protein, a higher lipid content, and nearly twice the caloric content of acorns.

Even so, there are never enough hickory nuts to sustain an entire squirrel colony, and the squirrels soon have to resort to acorns. In fact, acorns make up 90 percent of a squirrel's annual diet. Acorns from red oaks (*Quercus rubra*) are their second choice, after hickory nuts; then chestnut oak (*Q. prinus*), and finally white oak (*Q. alba*). White oaks are by far the most common oak trees in Toronto, and the city's squirrels may be said to live on white-oak acorns. Unfortunately, white-oak acorns are the least nourishing of all, and the squirrels have to make up in bulk what they lack in nutrition. High Park has one of the last remaining black-oak (*Q. velutina*) savannahs in southern Ontario, and High Park's squirrels are grateful for it. Too grateful, according to Steven Apfelbaum, the ecologist hired in 1993 by the city's Parks and Recreation Department to find out why the savannah was dying. Apfelbaum found that the enormous black oaks growing along the park's sandy ridge were very old; many of them were already 100 years old in 1873, when John Howard bequeathed to the city the 147 hectares along the Humber River that now form the western half of High Park (Toronto purchased the eastern half five years later, from the estate of General Thomas Rideout, for $15,000). Since black oaks normally don't live much longer than 200 years, Apfelbaum figures the trees will be dead in another twenty-five or thirty years – and there are no seedlings sprouting up beneath them waiting to become replacement trees. He doesn't know exactly why the black oaks are heirless, but he suspects that squirrels are eating all the acorns. He also says the turf grass (i.e., sown lawn) beneath the trees may be too thick to allow acorns that fall on it to penetrate through to the soil. Personally, I vote for the latter explanation. Turf grass is deadly stuff: even if black-oak acorns did get through it, most of the rainwater that falls on turf grass runs off (plant ecologists call turf grass "green asphalt"). Besides, turf grass shouldn't be in High Park any-way; practically John Howard's only stipulation when he willed his land to the city was that the park remain in its nat-ural state. Turf grass ain't natural. Squirrels are. And studies show that grey squirrels bury far more acorns than they dig up and eat – about 70 percent more – and are generally con-sidered to benefit hardwood stands by acting as inadvertent reforesters. The kind of soil squirrels choose to bury acorns in is usually the kind of soil the tree itself would choose if any-one asked it: not too wet, or the acorn will rot; not too dry, or the squirrel may not be able to find it again by smell.

Burying nuts is itself a deeply buried instinct in squirrels. Lucia Jacobs, while a graduate student at Princeton Univer-sity looking around for a doctoral topic in 1986, decided to study the way grey squirrels gathered and cached their food supplies. She brought a bunch of new-born squirrels into her apartment in April, raised them in cages on milk and soft foods until they reached adolescence, then put them in a dirt-bottomed enclosure and gave them hazelnuts to see what they would do. To her amazement, they invariably would "pick up a hazelnut for the first time, search intently for a suit-able burying site, and then, with great zest, dig a hole, both paws flying, the nut firmly clenched between tiny teeth, with all the apparent confidence and success of a jaded park squir-rel burying its millionth peanut."

Under more natural circumstances, grey squirrels almost always strip the husk off a hickory nut before burying it – per-haps to save themselves a bit of energy in the dead of winter, when every calorie counts, or else to prevent other squirrels from finding it by sniffing out the aromatic husk. They do this by nibbling a small hole in the side of the husk, inserting their lower front teeth into the hole and peeling off the covering in long, thin strips. Acorns, on the other hand, are usually buried whole, shell and all. When they have carried the nut or acorn to a suitable burial site, they dig a small hole a few centimetres deep, lay the nut in the hole, and ram it firmly in place by pressing their front teeth against it and pushing with their hind legs. Then they quickly push the earth back into the hole and pat it flat with their forepaws. Still not satisfied, they very carefully replace the leaf-litter and twigs on top of the spot, leaf by leaf, twig by twig, until the hiding place is completely concealed. This, too, is instinctual, not learned. When Lucia Jacobs gave a nut to a captive young squirrel in a cage with a bare wooden floor, the squirrel "buried" the nut in full view in a corner and then waved its paws over it as if arranging imaginary leaves over a phantom hole. I think wild animals benefit from having a lot of their drudge work programmed into them as instincts; it leaves their minds free to think of other, more important things. Such as flying.

In Ojibway legend, as the sun rose one morning it suddenly disappeared from view. All the animals went looking for it, and it was finally found by Squirrel; it had became entangled in the upper branches of a tall tree. Squirrel tried to free the sun by pushing it, but the sun was so hot it forced Squirrel down to the ground to cool off. This happened three times: the first time, Squirrel lost the fur on his tail. The second time his fur was burned black and he was almost blinded. The third time the sun's heat stretched Squirrel's fingers and joined them to the folds in his skin, but he finally succeeded in freeing the sun. In gratitude, the sun granted Squirrel anything he wanted. Squirrel thought for a moment, and then replied: I want to fly.

I think the squirrel at our cabin definitely wanted to fly too.

RECOMMENDED
READING

---

1. About Toronto and Ontario

Alex, J.F., and Switzer, C.M. *Ontario Weeds: Descriptions, Illustrations and Keys to Their Identification.* Ministry of Agriculture and Food Publication #640, 1976.

Chapman, L.J., and Putnam, D.F. *The Physiography of Southern Ontario.* Toronto, University of Toronto Press, 1951.

Faull, J. H., ed. *The Natural History of the Toronto Region.* Toronto, The Canadian Institute, 1913.

Fothergill, Charles. *Notes on the Natural History of Eastern Canada, 1816-1837.* Toronto, University of Toronto Press, 1934.

Gaymer, Rosemary. *Two in the Bush.* Scarborough, Consolidated Amethyst Communications, 1982.

Goodwin, Clive E. *A Bird Finding Guide to the Toronto Region.* Toronto, Toronto Field Naturalists, 1979.

Gosse, P. H. *The Canadian Naturalist.* London, John Van Voorst, 1840; reprinted in Coles Canadiana Collection, 1971.

Gregory, Dan, and MacKenzie, Roderick. *Toronto's Backyard: A Guide to Selected Nature Walks.* Vancouver, Douglas & McIntyre Ltd., 1986.

Herzberg, Louise, and Juhola, Helen. *Todmorden Mills: A Human and Natural History.* Toronto, Toronto Field Naturalists, 1987.

Howitt, J. Eaton. *Weeds of Ontario.* Ottawa, Department of Agriculture, 1911.

Ivy, Bill. *A Little Wilderness: The Natural History of Toronto.* Toronto, Oxford University Press, 1983.

Johnson, Bob. *Familiar Amphibians and Reptiles of Ontario.* Toronto, Natural Heritage/ Natural History Inc., 1989.

Johnson, Bob. *Amphibians and Reptiles in Metropolitan Toronto.* Toronto, Toronto Field Naturalists, 1983.

Johnson, Pamela. *The Ontario Naturalized Garden.* Vancouver, Whitecap Books, 1995.

Judd, William Wallace. *More Naturalists and Their Work in Southern Ontario.* London, Hearn/Kelly, 1892.

Judd, William Wallace, and Speirs, Murray, eds. *A Naturalist's Guide to Ontario.* Toronto, University of Toronto Press for Federation of Ontario Naturalists, 1964.

Logier, E.B.S. *The Snakes of Ontario.* Toronto, University of Toronto Press, 1958.

McNicholl, Martin K., and Cranmer-Byng, John L., eds. *Ornithology in Ontario.* Ontario Field Ornithologists Special Publication No. 1, 1994.

Mitchell, Margaret. *The Passenger Pigeon in Ontario.* Toronto, University of Toronto Press, 1935.

Montgomery, Frederick Howard. *Common Weeds of Ontario.* Toronto, Ontario Ministry of Agriculture and Food, 1959.

Sauriol, Charles. *Remembering the Don.* Scarborough, Consolidated Amethyst Communications, 1981.

Soper, James H., and Heimburger, Margaret L. *Shrubs of Ontario.* Toronto, The Royal Ontario Musuem, 1982.

Theberge, John T., ed. *Legacy: The Natural History of Ontario.* Toronto, McClelland & Stewart, 1990.

Ure, G. B. *The Handbook of Toronto, Containing Its Climate, Geology, Natural History, and Educational Institutions.* Toronto, Lovell & Gibson, 1858.

Urquhart, F.A. "The Introduction of the Termite Into Ontario," *Canadian Entomology.* Vol. 85, No. 8, 1953.

2. About Canada

Fenton, Brock. *Just Bats.* Toronto, University of Toronto Press, 1983.

Forsyth, Adrian. *Mammals of the Canadian Wild.* Camden East, Camden House Publishing Ltd., 1985.

Froom, Barbara. *Amphibians of Canada.* Toronto, McClelland & Stewart, 1982.

Hosie, R.C. *Native Trees of Canada.* Markham, Fitzhenry & Whiteside Ltd., 1990.

Lawrence, R.D. *The Natural History of Canada.* Toronto, Key Porter Books Ltd., 1988.

Frère Marie-Victorin. *Flore Laurentienne.* Montreal, Imprimerie de la Salle, 1935.

Wood, J.M., Dang, P.T., and Ellis, R.A. *The Mosquitoes of Canada, Diptera: Culicidae.* Ottawa, Agriculture Canada, 1979.

Wooding, Frederick H. *Wild Mammals of Canada.* Toronto, McGraw-Hill Ryerson Ltd., 1982.

Woods, S.E., Jr. *The Squirrels of Canada.* Ottawa: The National Museum of Natural Sciences, 1980.

3. Books on Related Topics

Balantine, Bill. *Nobody Loves a Cockroach.* Toronto, New York, Little, Brown & Co., 1967.

Evans, Howard Ensign. *Life on a Little-Known Planet: A Biologist's View of Insects and Their World*. New York, Dutton, 1968.

Forsyth, Adrian. *A Natural History of Sex: The Ecology and Evolution of Sexual Behaviour*. New York, Charles Scribner's Sons, 1986.

Garber, Steven D. *The Urban Naturalist*. New York, John Wiley & Sons, 1987.

Gibbons, Felton, and Strom, Deborah. *Neighbors to the Birds: A History of Birdwatching in America*. New York, W.W. Norton & Company, 1988.

Headstrom, Richard. *Nature in Miniature*. New York, Alfred A. Knopf, 1968.

Hubbell, Sue. *Broadsides from the Other Orders: A Book of Bugs*. New York, Random House,1993.

Jordan, William. *Divorce Among the Gulls: An Uncommon Look at Human Nature*. New York, HarperCollins, 1991.

Kennedy, Desmond. *Living Things We Love to Hate*. Vancouver, Whitecap Books, 1992.

McKibben, Bill. *The End of Nature*. New York, Doubleday, 1989.

Montgomery, Sy. *Nature's Everyday Mysteries: A Field Guide to the World in Your Backyard*. Shelburne, Chapters Publishing, 1993.

Pollan, Michael. *Second Nature: A Gardener's Education*. New York: Delta (Bantam Doubleday Dell), 1991.

Passmore, John. *Man's Responsibility for Nature: Ecological Problems and Western Traditions*. London, Duckworth, 1974, 1980.

Sokolov, Raymond. *Why We Eat What We Eat: How the Encounter Between the New World and the Old Changed the Way Everyone on the Planet Eats*. New York, Simon & Schuster, 1991.

Trefil, James. *A Scientist in the City*. New York, Doubleday, 1994.

Wilson, Edward O. *Biophilia: The Human Bond with Other Species*. Cambridge, Harvard University Press, 1984.

INDEX

Sisyrinchium, 155
Skinner, B.F., 192
skunk rabies, 20, 32
skunk, 16, 64, 66
sky lark, 231
Smith, Capt. John, 168-69
Smith, L.H., 44, 46
smooth-billed ani, 19
snake, 55-67; anatomy, 58-59,
    61-62; diet, 61, 62, 63, 64, 244;
    habitat, 55, 62, 63, 64; fear of,
    56-57, 59; myths about, 56-57,
    62; mating habits, 65-66;
    Species: blind, 40; garter, 5, 57,
    63-65, 66-67; hognose, 60-61,
    62; milk, 62, 63; massasauga rat-
    tler, 7, 60; northern brown, 62-
    63; red-sided garter, 64-65,
    137-38; timber rattlesnake, 60
songbirds, 41, 45, 49, 51, 105,
    111, 125, 232, 238
songs, birds, 90, 92, 237
Soper, James, 157
sorghum, 149
Spallanzani, Lazzaro, 204-5
sparrow: chipping, 49; Eurasian
    tree, 42; house (English)
    sparrow, 9, 38-54, 91, 97, 101-
    2, 111, 230, 231, 232, 234;
    savannah, 67; song, 49; vesper,
    230; white-crowned, 90, 91;
    white-throated, 45, 67, 90, 91,
    113
Sparrow Clubs, 51, 52
the Spit. *See* Leslie Street Spit
squirrel, 135, 241-57
staghorn sumach, 156, 157
starling, 9, 40, 97, 229-40; adapt-
    ability, 236-37, 238-39; diet,

239-40; territoriality, 232;
    Species: European, 86, 229, 231
Stedman, Capt. Jared, 200
stinging nettles, 8
Storeria dekayi, 62
Strickland, Samuel, 155-56, 244,
    245
Strom, Deborah, 105
Sturnidae (family), 235
Sturnus vulgaris, 230
sugar cane, 149, 150
Surgeonor, Doug, 143-44
Sylvilagus floridanus, 213
sylvine rabies, 20
Syringa vulgaris, 10

T

Tamiasciurus hudsonicus, 247
Taverner, Percy A., 50, 188, 232-
    33
Tennent, John, 7
termite, 68-86; compared with
    ants, 74, 75; importation of, 84-
    85; new species, 82-85; Species:
    dry-wood, 82-84; powder-post,
    82-83; subterranean, 69
termite control, 7-78, 79-82, 85-
    86
termite infestations (Toronto),
    72, 73-74, 78, 83
termite swarming, 207
Termite Tips newsletter, 69, 84
tern, 105; Caspian, 5; common,
    120
Thamnophis sirtalis, 63
Thamnophis sirtalis parietalis, 64
Thamnophis sirtalis rudix, 137
Thompson, Samuel, 154

This book is set in Weiss, a typeface designed by
Emil Rudolf Weiss, based on Italian Renaissance types.
It was issued in 1926 by the Bauer Foundry in Frankfurt.
It is one of the earliest contemporary serif typefaces with
italics based on the Chancery style of writing. A most
unusual aspect of Weiss is that the vertical strokes
are wider at the top than the bottom.

Book design by Gordon Robertson